Freshwater Ecology

FRESHWATER ECOLOGY
PRINCIPLES AND APPLICATIONS

Michael Jeffries and Derek Mills

Belhaven Press
London and New York

First published in Great Britain in 1990 2nd impression 1991, 3rd impression 1993 by Belhaven Press (a division of Pinter Publishers)
25 Floral Street, London WC2E 9DS

British Library Cataloguing in Publication Data
A CIP catalogue record for this book is available from the British Library

ISBN 1-85293-021-7 (hbk)
1-85293-127-2 (ppr)

Library of Congress Cataloging-in-Publication Data
Jeffries, Michael.
Freshwater ecology : principles and applications / Michael Jeffries and Derek Mills.
p. cm.
ISBN 1-85293-021-7 : — ISBN 1-85293-127-2 (pbk.).
1. Freshwater ecology. 2. Man—Influence on nature. I. Mills, Derek Henry.
II. Title. III. Title: Fresh water ecology.
QH541.5.F7J44 1990
574.5′2632—dc20 90-327
CIP

Typeset by The Castlefield Press Ltd, Wellingborough, Northants
Printed and bound in Great Britain by Biddles Ltd, Guildford and King's Lynn

Contents

Preface

Ironically, the planet we inhabit is called Earth. What makes planet three of our solar system special is water.

The presence of water on other worlds has a patchy history from the imagined seas of the Moon, canals of Mars and fetid, steamy forests of Venus to more recent, and definite, exploration of the atmospheres of some planets and their satellites. The extraordinary landscapes of Mars are very suggestive of surface water processes long ago.

The Earth alone retains surface water. Most of it makes up the oceans but the whole mass is not fixed. Water evaporates from sea and land, vapour clouds move in the atmosphere and rain falls to refill the seas, icecaps and underground stores and to supply the lakes, rivers, ponds and streams. For all their familiar beauty and importance in our lives, freshwaters are only a tiny proportion of the total water on Earth. They are a unique habitat with distinct flora and fauna, yet intimately linked to the wider world around them and sensitive to changes in this landscape, especially those brought about by humans. This book is written to explore and understand the ecology of freshwaters with three main threads. Firstly, the characteristics of aquatic habitats and, crucially, how these depend on the surrounding land and its use. Secondly, the ecology of the plant and animal communities, their links to the physical and chemical world and interactions with other wildlife. Thirdly, how human activities can perturb and destroy the habitats, their plants and animals and how these effects can be alleviated.

Freshwater ecology is a vast subject. There are many superb, thorough and detailed textbooks covering particular aspects in depth, whole volumes devoted to subjects that can scarcely take up a page or two here. There are also many reports and books that contain practical advice for good environmental management and conservation. Throughout this book references are given to these major texts so that students can follow up the basic principles outlined here and explore areas that interest them in all their complexity. The examples cited in more detail are largely drawn from widely available journals, again so that students can delve deeper. Examples have generally been chosen to point to the characteristic aspects of the topics under discussion but also links with other patterns and processes, especially to link up ideas across chapters. Some of the practical conservation texts are perhaps less widely available but are worth the extra effort to track down. This is especially important as the links between the pure and applied sides of the science bear fruit and, most important of all, if we are to save our planet's unique freshwaters from the many dangers that threaten.

Michael Jeffries and Derek Mills
Edinburgh
September 1989

Acknowledgements

We gratefully acknowledge permissions to use figures and tables from Akademie–Verlag, fig 4.6; The American Society of Limnology and Oceanography, figs. 5.1, 8.2; The Atlantic Salmon Trust, fig. 9.4; Blackwell Scientific Publications, figs. 1.4, 1.7, 4.8, 5.2, 5.3, 5.4, 6.2, 6.5, 6.9, 7.6, 8.4, 9.2, 9.3, 10.1 and tables 3.2, 5.2, 5.3, 6.2, 7.2, 9.1, 9.3; Sir John Burnett, fig. 4.3; Cambridge University Press, fig. 7.2; Ecological Society of America, fig. 6.3; Elsevier Science Publishers (England) fig. 7.3 and Physical Sciences and Engineering Division (Amsterdam), fig. 7.5; Fish Farmer Magazine, figs. 11.3, 11.4; The Forestry Commission, fig. 10.2; Controller of Her Majesty's Stationery Office, fig. 8.3; Institute of Terrestrial Ecology, fig. 1.2; Kluwer Academic Publications, figs. 1.8, 6.4, 7.7, 7.8; Macmillan Magazines Ltd., from Nature, Vol. 320, pp 746–748 Copyright (c) 1986 Macmillan Magazines Ltd, fig. 6.4; The Nature Conservancy Council, figs. 2.5, 8.1; Nordic Society Oikos, figs. 6.4, 6.10, tables 5.4, 6.1; Pergamon Press, fig. 7.1, table 7.1; Dr M.J. Phillips and Dr. M.C.M. Beveridge, figs. 11.3, 11.4; E. Scheizerbart'sche Verlagsbuchhandlung (Archiv fur Hydrobiologie) figs. 7.4, 9.1; Dr John Solbé fig. 9.4.

For the use of photographs we are indebted to Dr A. Bailey-Watts (Plate 4); D. Downie (Plate 29); Professor M. Frenette (Plates 28 and 30); Glasgow Herald (Plate 33); R. McMichael (Plate 23); Dr K. Morris (Plates 11 and 19); The North of Scotland Hydro-electric Board (Plate 27); The Scottish Wildlife Trust (Plates 3, 31, and 32).

1 Water

An enchanted vision of a verdant lake or gurgling stream is a picture that many people conjure up when thinking of freshwaters. These habitats are easy to envisage compared to the grades of terrestrial vegetation or invisible marine currents. Aquatic habitats are discrete, part of the landscape, but clearly delineated. Rivers and streams dissect the land, dividing it up. Lakes and ponds are scattered islands. This apparent separateness is so marked that freshwater ecology exists as a field often removed from wider ecological science.

This image of freshwater as isolated with a peculiar biology all their own is false. All freshwaters, from the longest rivers and broadest lakes to seasonal streams and rain pools, are intimately linked to the world around them. The water arrives from outside and eventually will leave. In between, the water will influence the surrounding land, perhaps directly in floods or erosion or by altering local climates and through the animals and plants using it. The waters will respond to climate. Chemicals will be gleaned from the surrounding area or arrive in the rain. It is very important to appreciate the role of these links between aquatic habitats and the landscape. Natural waters reflect the biogeographical world around them and unnatural waters, those altered, degraded and sometimes completely lost due to our actions, are so because of these same links.

1.1 The hydrological cycle

All water on the planet is linked by one global cycle, the *hydrological cycle*. No aquatic habitats are entirely static. Some water is always being added, some lost and some remains. The overall balance may be markedly in favour of one of these, for example a temporary pool drying out or a stream rising due to meltwater. The cycle can be simplified into a general pattern shown in Fig. 1.1.

A gross global budget of the whole planet's water can be drawn up and Table 1.1 summarizes this huge cycle. Three important points should be

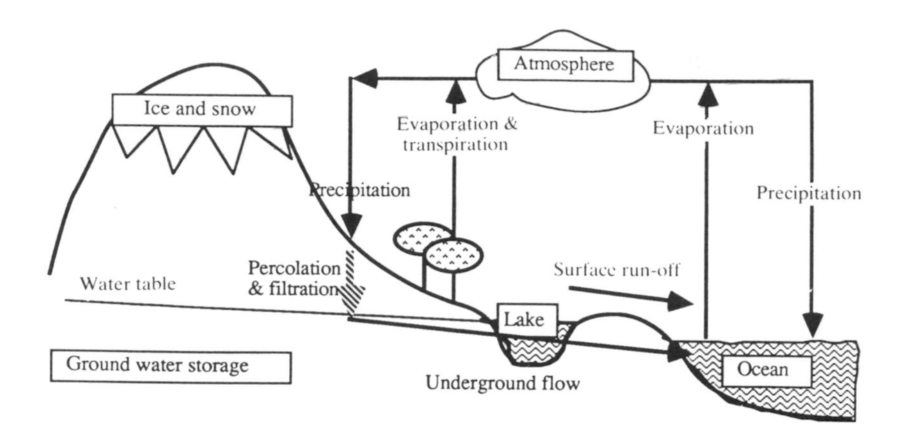

Fig. 1.1 The hydrological cycle. Evaporation of water from non-living sources is supplemented by water loss from plant life called evapotranspiration. This is important in terrestrial systems, and a major source of water loss from freshwaters with abundant emergent plant life.

Table 1.1 Estimated quantities of water in different parts of the hydrological cycle. Despite variations in estimates the same general pattern is evident; freshwater only makes up a tiny proportion of water on Earth, and genuine aquatic habitats such as rivers and lakes an even smaller amount

Component of	Volume (10^3km^3)	% of total water
Oceans	1 320 000–1 370 000	97.3
Freshwater		
Icesheets/glaciers	24 000–29 000	2.1
Atmosphere	13–14	0.001
Ground water (to 4000m)	4000–8000	0.6
Soil moisture	60–80	0.006
Rivers	1.2	0.000 09
Saline lakes	104	0.007
Freshwater lakes	125	0.009

noted. The links between lakes and rivers and the wider world are very evident from even such a simplified scheme. Secondly, the whole process is dynamic. Aquatic habitats are a link in a constantly changing bigger cycle. The water at each stage will react in different ways influencing subsequent links. Atmospheric water vapour, many tiny droplets with a huge surface area, are ideal to dissolve gases, thereby altering the chemistry of the water that then fills the lakes and rivers. Underground water, often heated or pressurized, dissolves minerals. Thirdly, the processes are essentially similar regardless of the size of the water body. Water is entering, remaining and leaving Lake Baikal, the largest freshwater lake on Earth, and is entering, remaining and leaving a village duck pond. Water enters, flows along and exits the River Amazon and in just the same way forms spatey streams on a mountainside. The scale at which ecological processes are measured and described may dramatically affect some conclusions. Differences in scale, both of time and physical space, should always be kept in mind in ecology. However, the general pattern of the global hydrological cycle is repeated at all scales in freshwater. Just as a gross global budget can be drawn up, estimates for a single aquatic system can be made.

Freshwater habitats may be vitally important parts of the landscape but they never make up much of the total area. An attempt to quantify the distribution of freshwaters in Britain estimated the percentage area occupied by freshwaters at between 0 and 1 per cent of the whole for most of England and Wales, with just a few highland areas with typically 2–5 per cent of their total area as freshwater (Fig. 1.2; Smith and Lyle, 1979).

1.2 Abiotic characteristics of water

As part of a vast geophysical cycle freshwater habitats are provided with water. Water is a very peculiar substance. It is the only liquid commonly found on the surface of the planet apart from occasional tar pits, molten rock, bubbling mud and the metal mercury, none of which is conducive to life. The strange properties of water create physical and chemical conditions to which wildlife must be adapted to exploit freshwaters. These physico-chemical constraints are abiotic (non-living) aspects of ecology and affect all aquatic habitats to some extent. However, the dominant influences vary from habitat to habitat: in rivers, the force of the water's flow may be of prime importance; in lakes, the depth to which light can penetrate may be most important. The same basic principles underlie all habitats and provide the basis for understanding how pollution and mismanagement can destroy systems. How these principles mould and form particular habitats, especially the differences between running and standing waters, will be explored in Chapters 2 and 3.

1.2.1. Physical characteristics of water

DENSITY

Freshwaters are waters in liquid phase, though many regularly solidify and all turn gaseous in part. Most living things also consist largely of water and so have a density very close to that of the medium itself. Living organisms will neither fall through water as they do through air nor, with the exception of tension effects, be able to live on it as they do on land. With little or no effort they can stay within the three-dimensional water column, passively

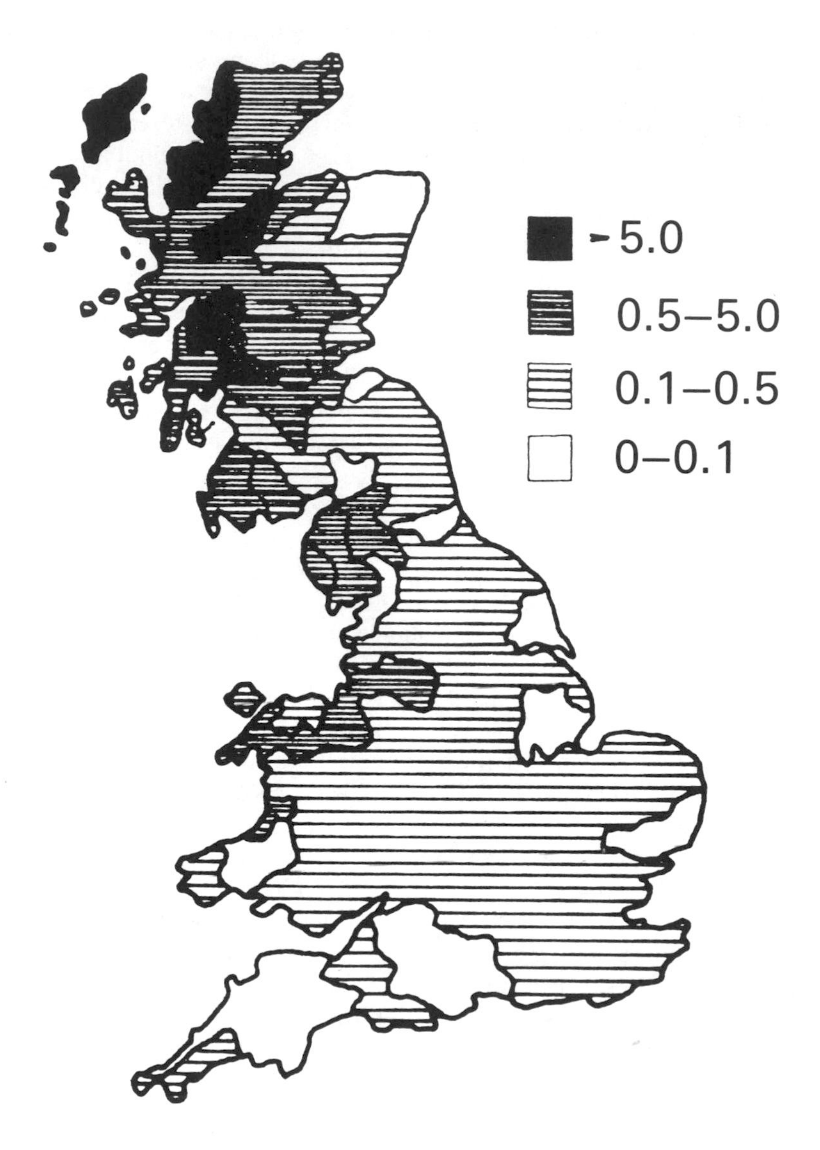

Fig. 1.2 The percentage area of standing water in Britain. (After Smith and Lyle, 1979).

floating and actively swimming but generally benefiting from the buoyancy. This is one unique feature of salt and freshwaters. Terrestrial habitats lack this exploitable matrix. Birds may fly but cannot spend all their lives aloft, unlike fish which can spend all their lives swimming. Many small insects and micro-organisms are blown into the atmosphere but they do not truly live in the air as do plankton in the open water column. Most plankton benefit by just floating passively in the water column. Many larger animals are able to maintain station by actively swimming, with the buoyancy as an added bonus to support their weight. Plants are also able to exploit an easier lifestyle. Many aquatic plants, relying on buoyancy for support, are flimsy compared to terrestrial forms. Free floating forms occur, some submerged (e.g. water soldier, *Stratiotes aloides* L.) and others on the surface (e.g. duckweed, *Lemna* spp. and water hyacinth, *Eichhornia crassipes* (Mart.) Solms-Laub).

This surface is an additional habitat created by water's peculiar physical structure (Fig 1.3). Water molecules have a marked affinity for each other, due to the electrochemical properties of individual molecules. At the air/water boundary they stick together so thoroughly that a surface tension is set up that will support light weights especially if they are spread out, for example floating leaves or the splayed legs of surface insects, and equipped with water-repellent devices (air-retaining hairs, waxy or oily chemicals). The tension hinders other life. Many aquatic insect juveniles have to emerge as flying adults and require adaptations to break the tension. Flying adults of beetles and bugs have to heave themselves clear. Taking air from above the surface also requires hydrophobic hairs or similar structures to break through the surface tension. The tension traps many terrestrial invertebrates that lack the repellent adaptations, providing a rich source of food (Mason and MacDonald, 1982).

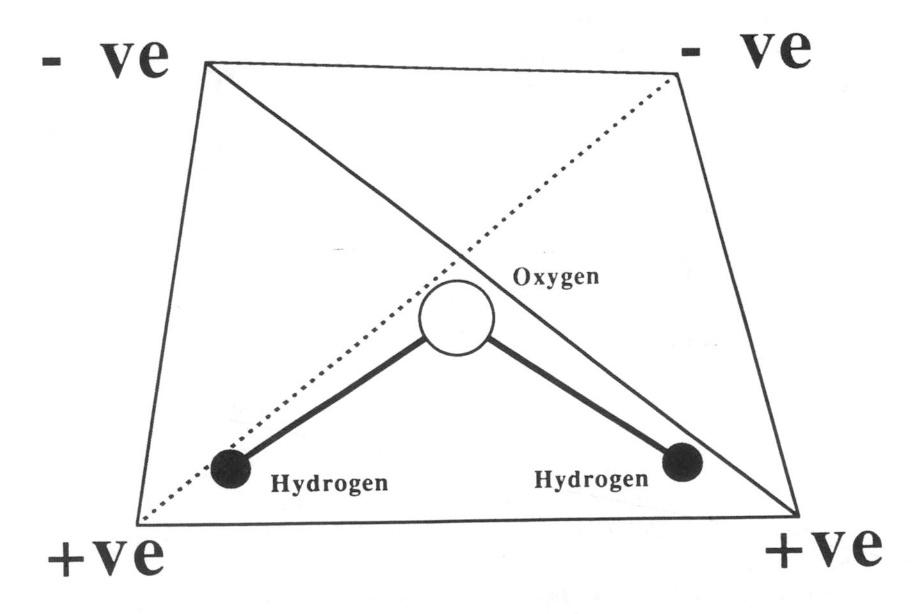

Fig 1.3 Stylized water molecule. The configuration of the oxygen and hydrogen atoms results in a slight positive (+ve) and negative (−ve) polarity in the molecules. The resulting electrochemical bonding gives water its unusual density and viscosity characteristics.

The density of the water column also creates problems. Water masses of different temperature differ in density. Water is densest at 3.95 °C. Water freezes to a solid at 0 °C and the fact that solid ice floats on liquid water would seem very anomalous if we were not so used to it. Solid ice has a regular molecular structure with slight gaps between the molecules. As ice melts, the regular form breaks up and these gaps fill. The density increases. This effect continues until 3.94 °C, so water at this temperature is densest due to the packing. Beyond 4 °C the increased energy of the warmed molecules causes more and more agitation. Above 4 °C the density decreases as the molecules increasingly separate. Their evaporation into a gaseous form is their ultimate fate with increasing heat. Between the extremes of ice and vapour density differences may occur between volumes of water. Currents of different temperature or layers in lakes differentially heated will not readily mix and their boundaries form barriers. To small plankton these may be physical, unable to push their way into denser water. Larger animals may avoid crossing between densities as conditions at one temperature may be poorer, perhaps the temperature itself or some difference in gas or nutrient solution.

The density of water gives the strength to the current in flowing systems. Confluent rivers do not always mix but continue side by side as distinct currents in the same channel. Movement of blocks of water of different density in lakes, as well as currents induced by wind action or rivers entering or leaving will also break up the apparent uniformity of standing waters.

Water even without the additional complexities of chemistry, habitat, plants and animals is a heterogeneous environment in which to live.

RADIATION

The sun's radiation sustains two important aspects of freshwater ecology. One is to heat the medium, resulting in density differences in the water mass, mixing and chemical alterations plus affecting the suitability of habitats depending on the temperature tolerances of the wildlife. The second aspect is providing light to power photosynthesis by plants, whether microscopic algae or macrophytes. Photosynthetically captured energy fuels their growth and eventually the whole community as the living plants or their dead remains are eaten. Visible light (which does not mean only that which is visible to us humans) has an important role as a natural clock governing the behaviours and life histories of many plants and animals. So the quality and quantity of light entering an aquatic habitat and how the water affects its availability are important.

The light penetrating water attenuates with depth. Different wavelengths of light are differentially absorbed. Distilled water, lacking all the dissolved chemicals and suspended particles which add to absorption, attentuates red wavelengths strongly, green and blue less so. However, no natural waters are so pure and organic particles markedly restrict blue light. Green is the least susceptible but also the least useful to most plants for photosynthesis; they reflect it, so appear green to our eyes (Fig 1.4).

At some depth the energy harnessed by photosynthesis will only just equal the respiratory requirements of the plants. This depth is called the compensation point, generally represented by the lower limit of photosynthetic algae in a lake, though the compensation point for any individual species can also be measured. Above this depth there is enough light to sustain net growth and this is called the euphotic zone. Below is the profundal zone. The precise depth of the compensation point varies with seasons and changes in water colour and cloudiness (generally termed turbidity). As a simple rule of thumb the compensation point occurs at the depth at which 1 per cent of the original incident light penetrates (Fig. 1.5).

Calculations of the precise euphotic zone depth are possible but a simpler and commonly used measure of the turbidity uses the Secchi disc. This is a white disc (sometimes chequered black and white) lowered into the water

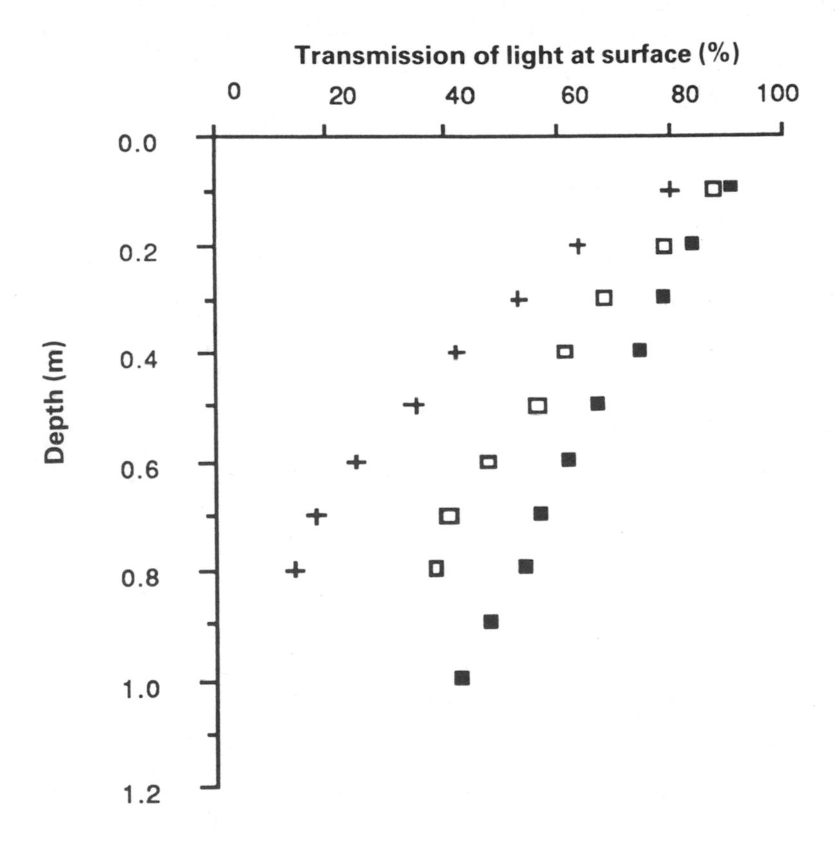

Fig 1.4 Transmission of light through the aquatic habitats: ■, Lake Windermere during an algal bloom; □, River Frome, a southern English river; +, effluent from a sewage treatment plant. (After Westlake, 1966)

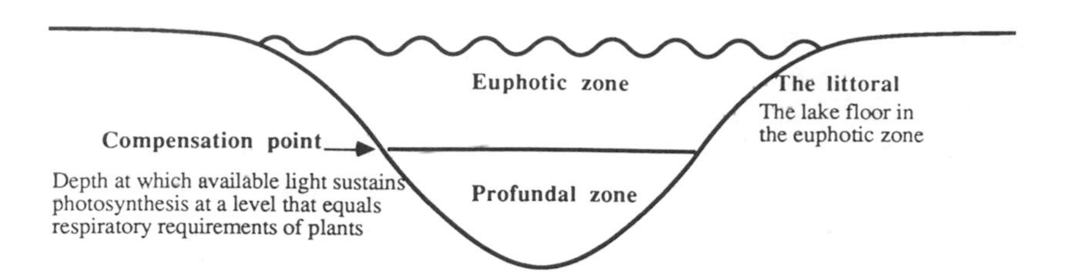

Fig. 1.5 Features of lentic habitats based on light regime. The compensation point, which is the depth at which photosynthesis just compensates for respiration, is taken as the divide between the euphotic and profundal zones. For a particular lake this point is generally marked by algae that can survive in very low light. Compensation points for individual plant species can also be measured.

column until no longer visible. This depth is called the Secchi disc transparency light extinction coefficient. It is named after Professor Secchi, whose efforts to quantify the turbidity were sponsored by Commander Cialdi of the Papal Navy in the nineteenth century. Although prone to differences due to plate size and the operator's skill, plus being only a relative measure of turbidity not absolute light energy, it is none the less a quick and intuitively nice technique.

Light affects the behaviour of organisms. Planktonic algae respond to different light intensity. Some actively swim up or down, for example the dinoflagellate *Ceratium hirudinella* OFM, or perhaps rely on the formation of gas vacuoles, for example many blue–green algae. Algae, especially diatoms, that live on the bottom are mobile and adjust their depth in the substrate as light changes. Algae and macrophytes can even alter the position of the photosynthesizing chloroplasts inside their cells according to light regime. Many animals in the open water column migrate, generally up at night and down during the day, for example zooplankton. Day length affects the life history, development and reproduction of many organisms, relying on light as a seasonal cue, or perhaps to coordinate daily rhythms. Shading and light regimes affect distributions across habitats and on a microscale around individual rocks and plants. Productivity and the growth of plants can be grossly affected, presenting management possibilities by use of appropriate shading. Kirk (1983) provides a very detailed text on light in the aquatic environment and how plants respond.

Much of the absorbed energy becomes heat. Water has unusual heating and cooling properties. Water has the highest specific heat capacity of any natural material, meaning the amount of energy needed to raise a mass of water by 1 °C is greater than that needed to raise an equal mass of any other material by the same amount. Similarly, a mass of water releases more energy as it cools by 1 °C than an equal mass of any other material. From an ecological point of view the basic result is that freshwaters take longer to heat up, or to cool down, than terrestrial habitats and the extremes of temperature they reach are less. The speed of change is slower. The thermal regimes are generally more clement than other habitats, though the resulting density changes may create their own problems. The thermal regime of freshwaters will interact with that of the surrounding land, generally as an ameliorating influence. The thermal and light regime may also be important on a microscale. Van der Velde *et al.* (1985) describe the diurnal changes in the abundance of surface dwelling fly (Diptera) species, as the microclimate over lily-pads changes, with a succession of species arriving throughout the day each with its own preferred temperature.

Heat will affect many other chemical processes of which oxygen solubility is the most important, cooler water generally dissolving more oxygen than warmer.

Temperature, like light, is an important influence on the life histories of animals and plants. Organisms all have their own maximum and minimum tolerances and optimal ranges for survival. These may be different at varying points in their life histories. Human activities can cause cooling or warming, perturbing and sometimes completely destroying the natural wildlife.

Lake sediments absorb, retain and transmit heat. The regime changes across the seasons and may be out of step with the open water column, acting as a heat source in winter and a sink in summer.

1.2.2 Chemical characteristics of water

Water dissolves a vast variety of chemicals, in the clouds before actually reaching the aquatic habitats as rain, or in the soil deeper underground in aquifers or over land and drip through trees. The precise water chemistry may characterize some habitats – acidic lochs, soda lakes, nutrient-rich rivers, iron-rich springs – and ultimately determine which animals and plants can live there. Equally many taxa may have one key element as a major determinant of their survival and success, such as diatom algae which require silica for their cell walls, molluscs which need calcium for their sells. Besides this multitude of specific links between chemistry, habitat and wildlife a few vital chemical cycles underlie the biology of all freshwaters: oxygen, carbon dioxide and the closely linked acidity/alkalinity, key nutrients (nitrogen and phosphorus) and organic compounds.

OXYGEN

Oxygen will dissolve in water but the amount varies with temperature. At atmospheric pressure, the cooler the water the more oxygen that dissolves. With increased pressure, more oxygen dissolves. Besides these basic physical influences the oxygen in natural systems also varies with mixing, turbulence, photosynthesis and respiration.

The amount of oxygen that will dissolve in water of different temperatures is shown in Fig. 1.6 for normal atmospheric pressure. At lower pressure, for example higher altitude, this will be less. The amount shown is the theoretical maximum measured as an absolute quantity (mg l^{-1}). The mixing, turbulence and biological activity greatly alter this amount so that sometimes less, sometimes more, than the amount predicted from the basic temperature/pressure conditions dissolves. The amount actually present compared to the theoretical level is another measure of oxygen levels and is called the percentage saturation, given as a percentage of the theoretical level based on temperature/pressure. A third, related, measure is called the biological (or biochemical) oxygen demand (BOD). This is a measure of the amount of oxygen used up by the life in a water sample, generally at the standards of 20 °C, over five days, and is often used as a measure of organic pollution. High BOD measures are taken as an indication of enrichment, the elevated oxygen use reflecting increased microbial activity as the polluting debris is decomposed. Most natural waters range from BODs of 0.5 to 7.0 mg l^{-1} O_2. Elevated levels typically occur with organic pollution, for example treated sewage 3.0–50.0 mg l^{-1} crude sewage 200.0–800.0 mg l^{-1} and silage 60 000 mg l^{-1} O_2 (Hellawell, 1986). All three measures of absolute saturation, percentage saturation and BOD are commonly used, giving insights into different aspects of the oxygen regime.

Oxygen levels often show marked daily cycles as plants release oxygen during daylight, sometimes supersaturating (i.e. greater than 100 per cent saturation) the euphotic zone, then respire at night causing a deficit

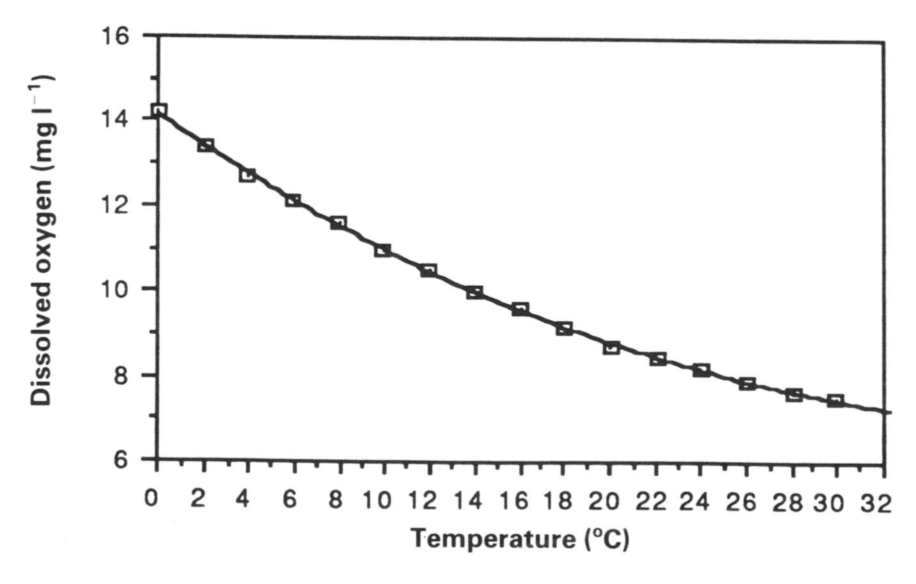

Fig. 1.6 Variation of dissolved oxygen with water temperature, at normal pressure, 1 atmosphere.

(Fig. 1.7). The availability of oxygen markedly affects other chemical cycles, especially once no oxygen is present when anoxic conditions alter the solubility of many other chemicals.

CARBON DIOXIDE AND ACIDITY

Carbon dioxide also dissolves readily and sustains a complex set of linked chemical processes especially those determining the acidity/alkalinity of the water and how this links to the carbon cycle and photosynthesis.

Carbon dioxide is more soluble in cooler water than in warm water and at higher pressures and much more soluble than oxygen (some 200 times). Hence, though the atmosphere contains much less CO_2 than oxygen (0.04% versus 21%) the CO_2 dissolves readily in the cloud vapour with its huge surface to volume ratio,

$$CO_2 + H_2O \rightarrow H_2CO_3$$

The latter is carbonic acid, a weak acid. The rain that falls on us, even in unpolluted areas, if any still exist, is not pure water but an acid. Carbonic

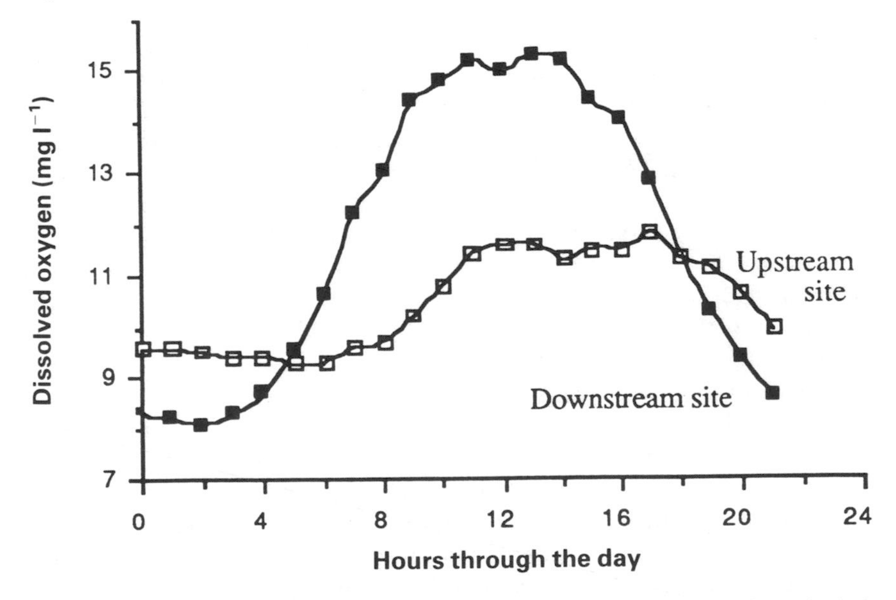

Fig. 1.7 Variation in dissolved oxygen at two sites on the River Ivel, in southern England. Site ■ is downstream of site □. Water takes 3.5 hours to move from the upstream to downstream site. The downstream data are shifted 3.5 hours backwards (i.e. over to the left) so that the difference between the curves at any specified time, represents the change in oxygen in a given body of water that has moved between the two stations. Diurnal variation and differences between sites are due to oxygen production by plant photosynthesis. (After Edwards and Owen, 1962)

acid dissociates to produce a free hydrogen ion (H^+) and a carbonate ion (HCO_3^-),

$$H_2CO_3 \rightarrow H^+ + HCO_3^-$$

The abundance of H^+ ions is what determines the acidity of a solution; the more there are the more acidic it is, though strictly speaking it is the activity of the H^+ ions that matters, not simply quantity. The acidity is measured as

$$\text{Logarithm}_{10} \frac{1}{\text{Amount of } H^+ \text{ ions present}} \quad \text{or} \quad -\log_{10}[H^+]$$

This measure is more familiar as the pH of the solution. For example, in pure water the H_2O molecules dissociate to a limited extent, so there are 10^{-7} mol l^{-1} of H^+ ions. The pH of pure water thus works out as

$$\text{Log}_{10} \frac{1}{10^{-7}} = 7 \quad \text{or} \quad \text{pH} = {}^{-}\log_{10}[H^+] = 7$$

Higher concentrations of H^+ ions (i.e. more acidic) give lower scores on the pH scale, for example H^+ concentrations 100 times greater than pure water, 10^{-5} mol l^{-1}, = pH 5. Lower concentrations of H^+ (less acidic) give higher scores, for example concentrations 100 less than pure water pH = 9. Waters of around pH 7 are described as circumneutral, those below pH 6 are acidic, those above pH 8 are alkaline. Carbonic acid dissociates readily and has a pH of 5.6. Natural rainwater has a pH of 4.5–5.6, the lower pH due to some natural pollution, e.g. volcanic emissions. Note that humankind has made a huge difference to the acidity of water in many parts of the world due to acid pollution and this is described in Chapter 8.

Other chemicals release or lock up H^+ ions to varying degrees, so the pH will alter as they are dissolved. However, a vital property of weak acids, such as carbonic, is that their dissociation is very malleable. Rain falling on areas naturally poor in free H^+ ions, for example alkaline geology such as limestone, will tend to dissociate more, releasing additional H^+ ions and compensating for the natural alkalinity. On areas of natural acidity, for example peat bogs where *Sphagnum* mosses lock up many alkaline ions, the dissociation is inhibited by the abundant H^+, again tending to compensate against the prevailing conditions. This tendency to compensate for the natural extremes is called buffering. Unperturbed freshwaters are generally well buffered, extremes in acidity or alkalinity are reduced. Some naturally acid systems are described in Chapter 8.

The bicarbonate ion (HCO_3^-) can further dissociate into $H^+ + CO_3^{2-}$, may react with water to produce a hydroxide ion, OH^- (e.g. $HCO_3^- + H_2O$

H_2CO_3 + OH^- and CO_3^{2-} + $H_2O \rightarrow HCO_3^-$ + OH^-). Bicarbonate, carbonate and hydroxide ions all contribute to the alkalinity of the water. Other chemicals, especially calcium, add to this measure which is essentially an overall measure of the chemicals present that shift pH to the alkaline side of neutral. Hardness is a related measure often used in water quality and refers to calcium and magnesium salts, combined with the bicarbonate/carbonate and other ions which compensate for the acidity. Either alkalinity or hardness is generally useful as a measure of the total acid-compensating ability of waters.

The acidity and alkalinity are major influences on the natural distributions of wildlife. The widespread impact of acidifying pollutants, 'acid rain', has become a colossal problem in some habitats.

The processes are further complicated as CO_2 is a primary component of photosynthesis. Aquatic plants, both algae and macrophytes, will use free CO_2 but many are able to utilize HCO_3^- ions. During daylight the uptake of CO_2 may exhaust supply and HCO_3^- is used. OH^- ions are secreted, replacing the HCO_3^-. Some of the freed CO_2 will precipitate as $CaCO_3$, a chalky deposit called marl. The result is an elevated pH (up to 9–10) within actively photosynthesizing weed beds, algal blooms and slimes on substrate. At night, with CO_2 released by respiration, the process reverses and pH can be lowered. These short-term diurnal cycles can be tolerated, but prolonged changes can alter water chemistry fatally and raised pH in large lakes and reservoirs with excessive algal blooms contributes to fish kills (Fig. 1.8).

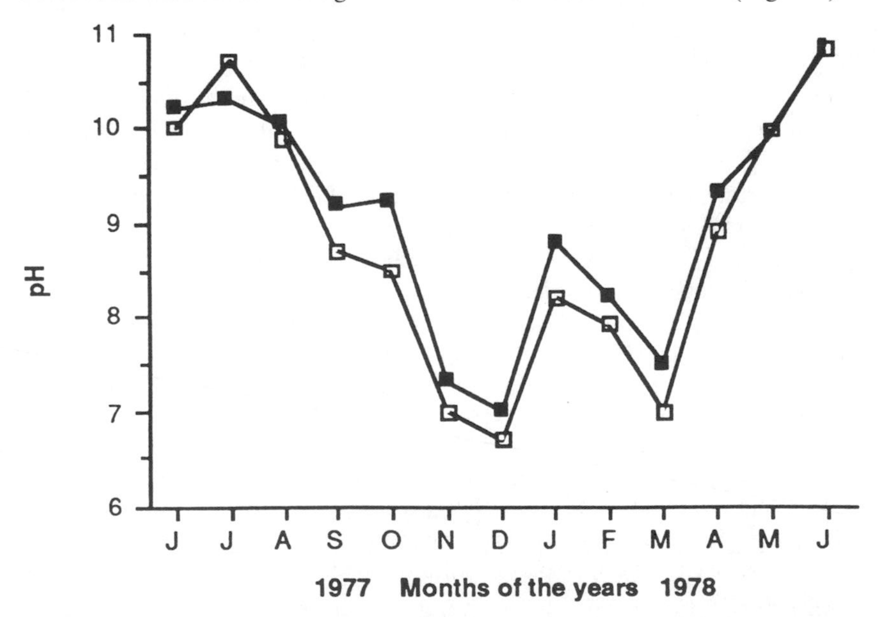

Fig. 1.8 Daily and seasonal variations in pH in a lake. The pH is elevated by photosynthetic activity, higher in summer and generally higher at midday than at night. ■, midday pH; □, midnight pH. (After Halstead and Tash, 1982)

NITROGEN, PHOSPHORUS AND THE TROPHIC STATUS

Plant growth relies on many elements to sustain health but two, nitrogen and phosphorus, are of special importance. The amounts required, much larger than those of other so-called micronutrients, coupled with the natural scarcity of available nitrogen and phosphorus, make them the two key nutrients. Their availability is a major factor in the plant productivity of a habitat.

Nitrogen is an abundant element at the Earth's surface: 80 per cent of air is nitrogen. Some organisms, a few bacteria and the related blue–green algae (Cyanophyta) can capture and use this raw nitrogen directly, a process called nitrogen fixing. Other life cannot and relies on nitrogen compounds such as ammonium (NH_4^+), nitrite (NO_2) and nitrate (NO_3) being available. These are created by electrical reactions such as lightning and released by nitrogen fixers and from organic matter, mostly the decaying detritus. The release from organic debris, especially the sediments, relies largely on bacteria that steadily oxidize the nitrogen from NH_3 to NO_3.

Phosphorus is a much scarcer element at the surface and does not occur as an atmospheric gas. It does not dissolve readily but will form insoluble complexes with common metals such as iron or calcium in aerobic conditions, settling in the sediments and becoming unavailable to plants. No organisms fix phosphorus. These problems conspire to make phosphorus the main factor usually limiting plant productivity even where nitrogen is plentiful. The ratio of nitrogen to phosphorus (N : P) may be a better indicator of likely productivity than simple absolute measures. Plants are adapted to make use of any phosphorus as it becomes available, gleaning and storing any excess to their immediate needs as luxury phosphorus. As with nitrogen, phosphorus occurs in many forms and is commonly measured as total phosphorus, phosphate (PO_4) and soluble reactive phosphorus.

These different compounds, like CO_2, O_2 and H^+, are components of big geochemical cycles involving the water, sediment, watershed, atmosphere and living things. Actual amounts present at an instant are often meaningless measures. Instead, understanding the overall budgets, for example the nitrogen budget, the phosphorus budget, and movement between components called the flux (e.g. the phosphorus flux), provides better understanding of the nutrient conditions in a habitat. The amounts of nutrients available over some time period, a rate rather than an instantaneous measure, are now commonly described and are termed the nutrient loading. The overall budget determines the productivity and this is often called the trophic, after the Greek word for feeding, status. A simple trophic classification is oligotrophic, mesotrophic and eutrophic, meaning little, middling and well fed. Nitrogen and phosphorus levels alone are often enough to characterize the trophic state, though precisely defined levels and the status they represent can be misleading. What constitutes a naturally nutrient-rich lake (eutrophic) in an area of barren, nutrient-poor geology might only represent levels of nutrient loadings that would qualify for oligo–mestrophic status in a rich, productive catchment. Trophic status for any one aquatic habitat is a condition relative to what the surrounding watershed, its topography and geology predict.

ORGANIC COMPOUNDS

Besides the inorganic carbon intimately tied to acidity and photosynthesis large amounts of carbon are bound up in organic compounds. These arise largely from decomposition of detritus but also some secretion, in particular by living plants. The smallest particles comprise the dissolved organic carbon (DOC), though this is partly colloidal, suspended rather than truly dissolved. Larger particles make up the particulate organic carbon (POC), sometimes combined with DOC in measures as total organic carbon (TOC). The two components differ more by an arbitrary measure of particle size than any profound biochemical factor. Very complex organic molecules such as amino acids, carbohydrates and humic compounds also occur and again divide into dissolved and particulate organic matter (DOM and POM): the latter sometimes splits further into coarse and fine (CPOM and FPOM). These intricate divisions have their uses, especially in describing the energy flow and trophic links in aquatic systems as different components are used by different organisms.

OTHER IONS, SALINITY AND OSMOTIC FACTORS

The variety of other ions dissolved in freshwaters all play some part, as trace elements are important as micronutrients and interact with the important cycles of oxygen, acidity and nutrients. Silica is especially important for its role in the algal plankton. Algae called diatoms (Bacillariophyceae) have cell walls formed of two overlapping shells (called frustules) made of silica. Their abundance and so that of competing species, is strongly affected by silica availability. Iron and sulphur are important in the oxidation and reduction chemistry, especially in the sediments. Some bacteria exploit the energy released by these chemical reactions to obtain their own energy needs. These are chemosynthetic bacteria. This energy harnessing is not efficient compared to photosynthesis. Chemosynthesis-based production does not support the teeming aerobic foodwebs that fill the verdant upper waters of lakes and rivers but in the dark, anoxic profundal gloom these bacteria may be the dominant life.

A measure of the overall quantity of dissolved substances is the conductivity of water to an electrical current. Distilled water has a conductivity of virtually zero, this increases to 400–600 microsiemens per centimetre in very solute-rich waters such as a chalk stream and to 1000–7000 in habitats such as coastal grazing marsh ditches. The conductance reflects the general levels of major ions, calcium, magnesium, sodium, potassium, chloride, very well.

These ions are also the dissolved components of salts so conductance is also some measure of salinity. Most freshwaters are, by definition, markedly less saline than seawater but some inland waters have a much higher salt content. They are hypersaline. They are not marine systems, they are not derived from cut-off seas nor are the fauna and flora marine, but they are saline inland waters. They can be small ponds such as the saline flashes that have developed over land subsidence in Cheshire, England, once a saltmining area, or huge lakes such as the famous soda lakes of Africa, where evaporation over many thousands of years has concentrated the solution of dissolved ions. Further examples are described by Williams (1987) and Melack (1988). The majority of freshwater animals and plants have their origins either in marine systems or have re-entered the water from terrestrial habitats. The biota must have adapted to the osmotic differences between freshwaters and their original environments. Sea water, which is similar to body fluid osmotic potential, allows lax osmotic regulation so marine organisms would need to develop effective osmoregulation. Terrestrial organisms, adapted to resist water loss would have to cope with a potential excess.

1.3 Time and seasons

Measurements of the physico-chemical characteristics of habitat are often instantaneous. These data tell us what the environment is like at a special instant on one day of one year of the whole life of a lake or river. Sometimes that simple information is perfectly useful and sampling programmes can be devised to take measures at intervals of minutes, days, weeks and years to build up a better picture. The chemical processes and physical conditions vary as part of natural cycles under the influence of season and climate, even 0as a result of the activities of the animals and plants.

To fathom the ecology of freshwaters a feel for the age of habitats and the changing conditions with time is important. This is simply another facet of the problem of scale in ecology. In Chapters 2 and 3 the lives of lakes and rivers will be described in more detail, how their physico-chemical make-up changes, how river channels and lake basins age. These are individual lives. More general patterns occur, rhythms playing on scales from the vast to the tiny. The whole planet may be subject to some regular cycles on a vast scale, from the impact of meteorites and global volcanic activity to climatic shifts of ice ages and warm periods. These colossal events will include freshwaters in their massive powers of destruction and creation. As factors responsible for the form and variety of freshwaters we see nowadays their effects are largely too general. However, some regions still show their influence, for example the Great Rift Valley lakes in Africa with their fish populations isolated for very many years or the glaciated landscapes of the Northern Hemisphere.

More evident to us and vital for the application of good management are the temporal patterns imposed by years (annual), months (lunar) and days (diurnal).

In areas with marked seasons, temperate winters and summers, tropical wet and dry, aquatic habitats show startling annual cycles. These may be to the extent of forming and disappearing (e.g. vast tracts of seasonally flooded tropical forest, tiny rain-fed ephemeral pools). The whole ecosystem starts up and is lost. Permanent waters will show changes in volume and radiation received. The latter is important, affecting heating and density and photosynthesis. The seasonal change in light acts as a clock governing development and metamorphosis of wildlife.

Lunar rhythms are also often related to the brightness of the moon, perhaps linked to tides where rivers and sea meet. Lunar cycles drive migrations such as those of the eel, *Anguilla anguilla* (L.). Daily cycles are more familiar. In large water bodies substantial changes in volume, density and heating are limited over such short periods but marked pulses of photosynthesis and the related oxygen and pH changes occur, followed by oxygen deficits at night. Small waters will show dramatic daily changes. Animals and plants respond to these rhythms with synchronized migration up and down in the water column; drifting in the flow of rivers and massed emergence are synchronized.

Time, whether evoked by the successional changes in a lake, the evolution of a river channel or the recycling patterns of habitat and life, is as important a factor as any of the more obvious physico-chemical parameters.

1.4 Water and watersheds

Just as time is easily overlooked as a thread in the pattern of life so the influence of the surrounding land on freshwaters can be neglected. The physico-chemical conditions in a habitat depend intimately on the adjacent land. The environs from which water drains into a lake or river is the catchment. The term 'watershed' may also be used, though in Europe this tends to mean the dividing line between neighbouring catchments rather than the catchment's drainage area itself.

The gross geography and geology will have a massive role. The shape (area and depth) of a lake will depend on the basin it fills, affecting heating, cooling, circulation and chemistry. The geology may determine the natural trophic status. Catchments alter the acidity, perhaps buffering, perhaps skewing the balance to one extreme or the other. Topography alters light regime, especially in high latitudes where the sun is seasonally variable, sometimes overhead but low in winter and obscured by hills. Rivers will be similarly affected with additional changes as they flow over rocks and sediments of different resistance and geology, as slopes change, as catchments meet and channels join.

The vital consequence of these natural links is that any human activity in a catchment is likely to affect the water sooner or later: agriculture, industry, forestry, building, impoundment even the transfer of water from one catchment to another. Effects can be immediate, for example sediment run-off from logging or roadworks, or take many years, for example nutrient build-up in underground aquifers. Nowadays, the effects are not even restricted to individual catchments. The global hydrological cycle allows some perturbations, notably artificial acidification of rainwater, to be an international problem. From the smallest pond to the biggest lake, freshwater habitats respond to the wider world around them in continual cycles of chemistry, form and time.

2 Standing Waters

The beauty of lakes and ponds, often a focal point in equally enchanting scenery, their histories and the uses to which they are put are captured in the variety of names: lakes, lochs, loughs, ponds, polls, meres, lochans, broads. While a particular title conjures up a definite picture, distinguishing each of these terms precisely is difficult. However, they are all similar in that they lack any dominant, unidirectional flow – the distinguishing feature of rivers and streams and their equal treasury of names. Standing and flowing waters are commonly termed lentic and lotic. Within each the essential physico-chemical properties of water and the progress of time create conditions as catchment and habitat interact and develop. This chapter outlines the characteristics that assume special importance in lentic waters, how they form and age, in response to their catchments. Eventually some of the finer differences associated with the many types of lentic habitats which we intuitively recognize in their many names are picked out.

2.1 Lakes

2.1.1 The nature of standing water

The language of science is harsher than folk names but the terminology given to different parts of a lake, to the water masses and wildlife is useful for precision and concise discussion. Lentic habitats contain areas of differing light, heat and physical structure and much of the useful jargon describes these zones.

The well-lit surface waters comprise the euphotic zone, their lower limit at the compensation point and below this is the profundal zone. Closely linked to light is heat radiation and the lack of a strong continual, undirectional current in lentic waters allows another of their important characteristics to evolve. The large amount of energy absorbed by water to heat it, coupled with rapid attentuation so that very little ever penetrates deep, results in a thermal stratification. A warm upper layer, the epilimnion, may form, floating on a cooler lower volume, the hypolimnion (Fig. 2.1). The transitional zone, often abrupt, is the metalimnion, though the term for the sharp temperature gradient across this zone, the thermocline, is commonly used synonymously (Fig 2.2). The development of stratification varies with geography and season but the same general terms apply. In seasonal climates stratification eventually breaks down, in temperate

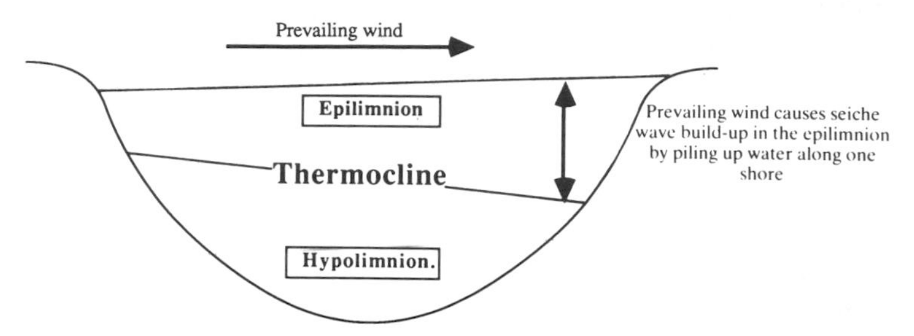

Fig. 2.1 Thermal stratification in a temperature lake in midsummer

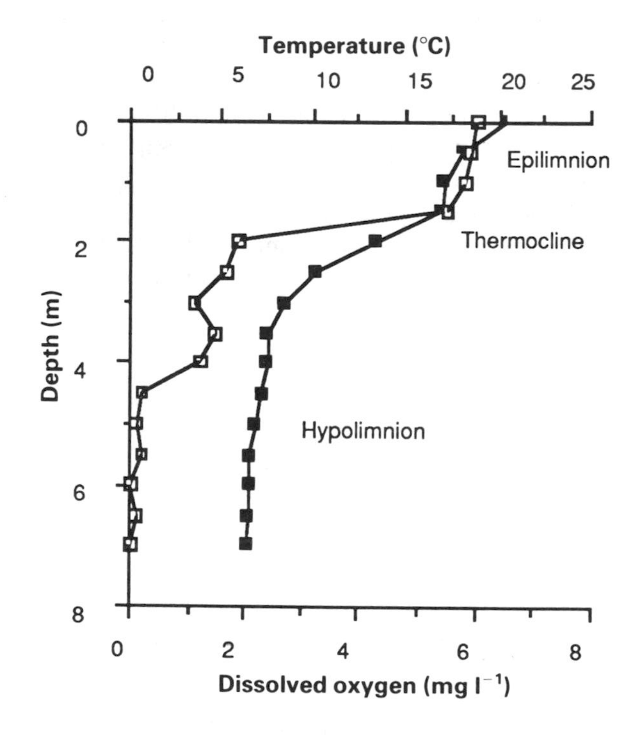

Fig.2.2 Summer stratification in one of the Uath Lochans, Glen Feshie, Scotland. This dystrophic lochan is sheltered by hills and forest allowing a warm epilimnion to develop in summer. Note how the stratification greatly affects the dissolved oxygen. Below the thermocline, although the water is cooler there is much less oxygen as it is used by decomposition and not replaced across the stratification barrier. ■, temperature; □, oxygen.

winters when radiation is low and gales mix the water. This is often referred to as the turnover and these stratifying–mixing lakes are called holomictic (the whole lot mixes). Very deep lakes may never completely mix (meromictic). In such cases the upper layer, which may have its own circulation is called a mixolimnion and the lower layer is the monimolimnion, typically very rich in dissolved ions, saline and anoxic. The high concentration of salts may be the result of solution from the sediment but in saline springs they also sustain such systems.

While the stratification remains intact the epilimnion and hypolimnion can operate as very discrete masses, not mixing. One common result is that prevailing surface winds push the top of the epilimnion to one shore of the lake. Here the water literally piles up, increasing epilimnion depth on this side with a consequent decrease on the opposite shore. The surface level on the shore with water piling up will be higher than on the lee shore. Once the wind dies this imbalance will see-saw back and forth, the epilimnion recovering, overshooting and rocking back and forth until momentum is lost or new wind factors take control. Such oscillations are seiches and are in effect gigantic, whole basin waves.

The basin's topography and water depths are important. The lake bed, from high water to the bottom of the euphotic zone is called the littoral. In the profundal, the bed is called the sublittoral, or profundal, the same as the dark water mass overlying it. The littoral zone is commonly covered by macrophytes and their different growth forms (emergent, floating leaved, submerged) reflect different depths. The open water above the sediments and outside of the weed beds is the pelagic zone, which embraces the open water across the euphotic/profundal and epilimnion/hypolimnion divides. These terms can be used to describe the animals and plants too, for example the pelagic community or littoral community. Some useful terms apply more specifically to the wildlife rather than to the physical environment. The pelagic community consists of large and small organisms. Those swimming freely and that can move relatively easily between zones are the nekton, largely fish. Smaller organisms, perhaps able to swim but largely at the mercy of water movements are plankton: zooplankton (animal) and phytoplankton (plant, though fungi and bacteria often are included). The term seston encompasses both living plankton and dead suspended debris. Animals living on the sediment are the benthos (hence benthic community). The community within the sediments is the psammon, though this properly refers to sand and other fine particle substrate. Many microscopic beasts live their lives entirely within the water-filled cavities, the pores, between sediment grains. This is the interstitial community (Fig. 2.3).

Life associated with the immediate surface is the neuston. That living on top the epineuston, that immediately below the hyponeuston.

With time, lentic bodies tend to fill in with deposits from their catchments and organic production of their own. They are essentially depositing habitats. Lotic systems are generally erosive, increasing their length with

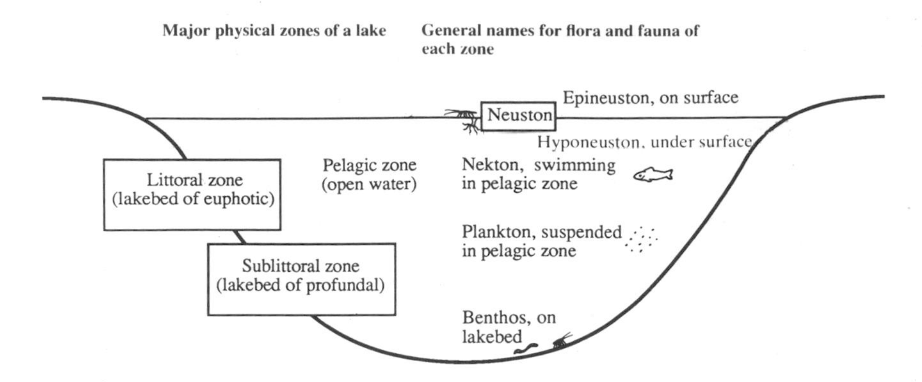

Fig. 2.3 Major physical zones of a lake and the general names for flora and fauna in or on these areas.

Plate 1 A natural dystrophic lochan, Uath Lochans, Scotland. The waters are naturally acidified and fringed by encroaching *Sphagnum* moss (see Chapter 8). (Photo: Michael Jeffries)

time, though some stretches are dominated by deposition. As lentic habitats fill so the extent of different habitats within them alters with the littoral zone encroaching as plants spread inwards with species eventually properly characteristic of wetland rather than true open water dominating. The zones of plants, from submerged, floating leaved, emergent and the following wetland vegetation are a succession each stage being a sere, (hydrosere to be precise since they refer to aquatic habitat). The ultimate fate of all lentic systems, the end of the succession, is a fully terrestrial flora on the infilled site. The variety of wetlands that arise in between (marshes, swamps, mires, bogs, fens, again the evocative names seem endless) are important habitats in their own right. However, apart from a brief reconnaissance later, this book will stop at the emergent phase of succession when the habitat is still primarily aquatic.

2.1.2 The formation of lakes and ponds

Lentic systems will form in any depression or hollow that can, even briefly, hold water. Tiny pools in between the leaves and stems of tropical bromeliads, in pitcher plants (these plant ponds are called phytotelmata; Frank and Lounibos, 1983), in the bowls of trees, in discarded tins, tyres, even the vases for flowers in a cemetery (Barrera-Rodriguez *et al.*, 1982) provide habitats with their own special fauna (Williams, 1987).. The vast mass of freshwater is contained in large lakes and they are also vastly longer lived habitats. Their formation, resulting shape and surrounding landscapes, vary greatly, and so do their abiotic features and inhabitants. Many efforts have been made to classify lake formation, resulting in sometimes baroque detail (Hutchinson, 1975). There are three general causes with specific subdivisions.

1. *Rock basins.* Formation of a lake basin by depression of the landscape resulting from:
   (a) Volcanic activity, in the crater of collapsed dome (caldera). e.g. Eifel Maare, West Germany.
   (b) Glacial activity, gouging out a depression or in basins left in glacial deposits. e.g. Lake District, U.K.
   (c) Tectonic activity, such as the lakes in the African Rift Valley.
   (d) Solution of the bedrock, dissolving out a basin. e.g. Süssersee, East Germany.
   (e) Meteorite impacts (perhaps the longer rhythms of time are not so invisible after all). e.g. Lago di Monterosi, Italy.

2. *Barrier basins.* Formation by the imposition of a barrier across a previously open channel.
   (a) Landslides. e.g. Lake Tarli Karng, Australia.
   (b) Lava flows. e.g. Lake Okataina, New Zealand (flows down a caldera)
   (c) Windblown deposits.
   (d) Glacial ice or moraines. e.g. Mürjelensee (ice) and Lago di Muzzuna (moraine), both in Switzerland.
   (e) Coastal systems, such as spits and dunes. e.g. Slapton Lee, U.K.

   These barriers often close a river channel, the natural dam forming a lake. River channels themselves can spawn lentic habitats, notably when meanders and other channels are isolated from the main channel or sudden loss of slope results in an ill-defined channel, open water spreading out across a wetland. Rivers that seasonally separate into isolated pools also briefly function as lentic habitat.

3. *Organic basins.* These are created by the action of living things, e.g. beaver dams, and differential vegetation growth. With the advent of human activity this form of habitat creation has become very important. Certain waters, such as small ponds in lowland agricultural zones of Europe are almost exclusively artificial. Massive impoundment schemes have created huge lakes, often on previously large river systems.

2.1.3 The biography of lakes

Once formed, lentic waters develop and change greatly with time. The precise life history of an ephemeral pool or tin-can puddle and that of a huge lake are different both in what happens and the time scale, but most ponds and lakes follow a general pattern to final extinction. There is a fine difference between the classical and the clichéd example but an outline of the general changes, the biography of a typical lake, would run as follows.

Plate 2 Loch Scionascaig (sheenaskayg), north-west Sutherland, Scotland. A typical oligotrophic loch with steep surrounding terrain and practically no littoral zone. The fish fauna comprises Arctic charr and brown trout and is typical of many large Scottish oligotrophic lochs. (Photo: Michael Jeffries)

Newly formed lakes are generally barren of life. Few plants and animals will have established themselves, sediment may be limited and nutrients scarce, certainly in naturally created basins. The scarcity may be relative, lakes on nutrient-rich geology start out relatively barren compared to their final condition, but many lakes are poor in absolute terms. This is partly an effect of time and place. The majority of land on this planet is in the Northern Hemisphere and was extensively glaciated in recent geological times. The combination of glacial removal of soil and cool climates conspires to make the multitude of glacially created lakes barren at birth. It will take time for animals and plants to colonize. The impoverishment of nutrients and life causes low productivity. Lack of sediment and extensive weed beds also reduce available niches. The supply of nutrients is dominated by input from outside of the lake itself, even if that supply is feeble. Such sources are called allochthonous. Production within the habitat itself is called autochthonous. Deep, young lakes also have a relatively large hypolimnion to epilimnion. This restricts nutrient cycling, preventing recirculation to the epilimnion and also, since there is little detritus to decay, they remain aerobic, which inhibits the solution of phosphorus. The increase in autochthonous algal and macrophyte production will be extremely slow. Initial predominance of phytoplankton production, over the macrophytes which lack substrate and colonize slowly, may scarcely match losses especially if there is an outflow river. In large, deep lakes with a limited littoral zone relative to overall volume, the build-up of organic deposits will remain slow even once macrophytes establish. Alternatively, shallow lakes and ponds, with a large, productive littoral may show a rapid acceleration in autochthonous production. Once even a large lake reaches this terminal stage with organic production large compared to losses and overall volume, the final infilling and transition to wetland is extremely rapid.

Across much of the Earth lakes are largely postglacial features but a few, generally in the centres of ancient continents, are much older. Lake Eyre, in Australia, dates back to the Cretaceous period (De Dekker and Williams, 1986) and Lake Chad, in Africa, may be even older (Beadle, 1974). The ancient lakes of the African Rift area are more recent, for example Lake Victoria dates from 25 000–30 000 years ago and is still growing (Burgis and Morris, 1987).

Increasing input of nitrogen and phosphorus will increase the productivity of the lake. Simple sample measures may reveal no more nutrients dissolved at any instant than in barren waters, but this is because the plants so effectively take them up as soon as they become available. The differences show in the organic productivity. The changes may be obvious in the steady encroachment of macrophyte seres as sedimentation fills the basin. The nutrients are now more abundant, the lake is becoming eutrophic. Eventually emergent seres dominated by dense stands of reeds and sedges smother the remaining open water. The wetland increasingly dries out, evapotranspiration from the plants themselves adding to the rate, and woody shrubs invade. The variety of wetlands that form depends on the soil, climate and hydrology just as the original lake did, and are as diverse.

Plate 3 Knowetop Loch, Dumfries and Galloway, south-west Scotland. A mesotrophic loch with a wide littoral area and surrounded by relatively low-lying farmland. (Photo: Scottish Wildlife Trust)

Plate 4 Loch Leven, Kinross-shire, Scotland. A well-known Scottish eutrophic loch situated among rich low-lying arable farmland and small townships, all of which contribute nutrients to the loch in the form of fertilizers, farm animal waste and sewage. This water was intensively studied as part of the International Biological Programme. (Photo: A. Bailey-Watts)

2.1.4 Lake classification

Lakes can be described and classified in many ways, choice depending on the reason for wanting to do this in the first place: for example by chemistry, physical form, size, geographical position, depth, phytoplankton, zooplankton, macrophytes, invertebrates, fish, human use.

Trophic status and thermal stratification are two criteria regularly used. Lakes are often described as either oligo-, meso- or eutrophic reflecting productivity and in turn echoing age and catchment. Classifications with quantified limits to each category have been devised but remember that absolute measures can be very different between waters but as a reflection of their age, catchment and any pollution they could still be assigned to the same category.

The thermal regime and resulting stratification is another scheme which reflects geography and morphology. Some lakes in the Arctic and Antarctic plus Alpine area never stratify or mix. Iced over all year they are very constant, never warmed enough to separate, never stirred. They are called amictic (don't mix). Some lakes stratify once a year and mix once a year. These are monomictic (mix once) and are subdivided further into warm or cold monomictic. Cold monomictic do not warm above 4 °C. They ice in winter (the stratification) and mix in summer when the ice cover melts. Warm monomictics are always warmer than 4 °C, stratify in summer and mix in winter. Lakes that stratify twice a year (summer epilimnion/hypolimnion, winter ice cover) with mixing twice a year in between are called dimictic. The lakes of northern temperate latitudes are typically dimictic. In the tropics polymictic and oligomictic lakes occur. Polymictic (many mixed) show such limited thermal gradients that they stratify and mix more or less every day. No long-term stable stratification occurs. Oligomictic lakes are deep and stable with continual stratification. The long-term separation of upper and lower water masses results in a hypolimnion that is so steeped in dissolved minerals that it is often refered to as a monimolimnion.

Plants have been used to describe types of lakes. The original use of the terms oligo- and eutrophic, to define lake nutrient status, came before accurate physico-chemical measures could be used and was based in part on the algal community present. A more refined version was suggested by Nygaard (1949), called the phytoplankton quotient. This is essentially the ratio of species typical of eutrophic conditions to those of oligotrophic conditions, notably the desmids (Desmidiaceae). Brooks (1964) gives an example for some Scottish lochs with the quotient defined as:

$$\text{The phytoplankton quotient} = \frac{\text{Cyanophyceae} + \text{Chlorophyceae} + \text{Centrales} + \text{Euglenineae}}{\text{Desmidiaceae}}$$

In this example a good correlation was found between the quotient and physico-chemical measures, with quotients of greater than 2.0 rating as eutrophic, of less than 0.8 as oligotrophic, and those between, mesotrophic (Fig. 2.4).

Macrophytes have also been used, especially for conservation work in Britain. The Nature Conservancy Council bases a lot of its freshwater surveys on macrophytes, analysing community patterns to look for typical types of lakes (Fig 2.5; Palmer, 1989). Palmer produces a national classification for lakes in Britain, with the macrophytes generally reflecting the trophic status of the waters, which in turn is due to biogeographical features and also human activities. This scheme allows natural characteristics to be integrated with possible evidence for pollution damage.

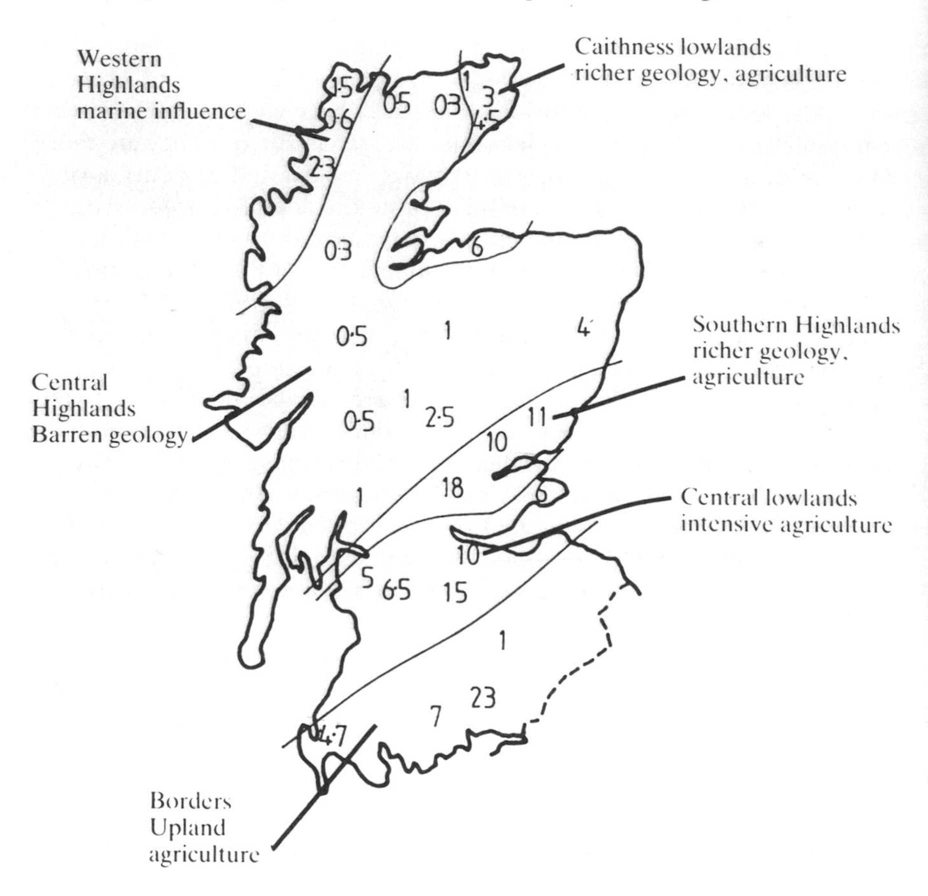

Fig 2.4 Use of phytoplankton to classify lakes. Brook (1964) classified Scottish lochs using a phytoplankton quotient. This is essentially a ratio of nutrient-rich water taxa to nutrient-poor water taxa. The higher the quotient, the richer the trophic status of the loch. Note that the results reflect general geology, location and land use.

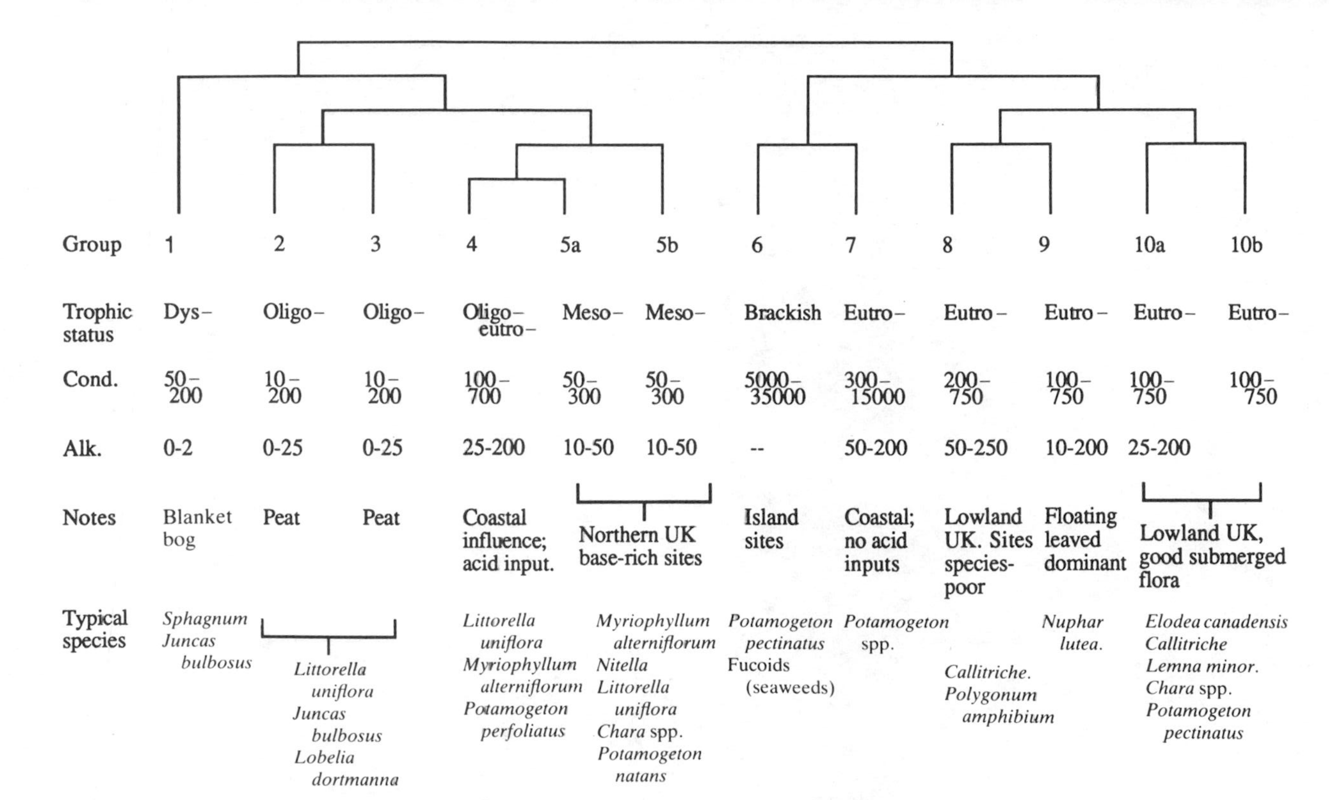

Fig. 2.5 Classification of 1124 lakes throughout Britain, based on macrophytes. Cond. = conductivity (umhos). Alk = Alkalinity ($CaCO_3$, mg l^{-1}). (After Palmer, 1989)

Plate 5 Lake Myvatn, a subartic eutrophic lake in north-east Iceland. Named midge lake, because of its vast chironomid production, it is known worldwide for its large waterfowl population. A local company dredges the deposits of diatomite from the lake for the manufacture of kieselguhr. It is feared by some that the removal of so much silica, which is an element required by diatoms, might have an effect on the food chains in the lake and affect the breeding success of certain duck species which feed on the chironomids and cladocerans. A monograph on the lake (edited by Jonasson, 1979) covers all its unique aspects. (Photo: Derek Mills)

2.2 Ponds

Despite the fact that everyone knows what is a pond and what is a lake it is impossible to provide a precise, technical difference. Some authors (Wetzel, 1975) describe 'pond' as simply an old term for shallow water bodies that are none the less lakes. The later stages of a lake as it fills in share many of the abiotic characters of a pond, and much of the wildlife, but ponds differ sufficiently to make the distinction useful, not least for the practical purposes of conservation, given the anthropogenic origin of so many of them. These specific characters are:

1. Many are artificially made and do not have the long history of a larger lake underlying their current form.
2. Their life history and characteristics have a speed and intensity that allow factors other than the dominating pace of lake productivity to be

Plate 6 A Swiss alpine lake, Schwarzsee, lying at 2580 m above sea level on the lower slopes of the Matterhorn above Zermatt. Although it is covered by ice for a large part of the year, it has a resident population of minnows (*Phoxinus phoxinus* L.). (Photo: Derek Mills)

dominant. In particular, their huge number but small size results in variable colonization and annual perturbation causing very variable communities.

3. Many small ponds are transitory (on an annual scale) ephemeral habitats with an unusual, specific biota that does not occur in lakes.

Naturally created ponds show the same general history as lakes. However, unless perturbed or polluted their small volume and shallowness result in very rapid progress to the point where autochthonous production dominates filling up the basin. Many ponds are entirely artificial and their original purposes diverse: for example livestock waterholes, fire-fighting, fish farming, duck rearing, shooting, osier beds, ornamental, defensive, sporting, reservoirs, mill ponds, effluent settling, mining, bomb craters. Ponds are short-lived landscape features, often lasting only a few hundred years. Their survival and precise form depend largely on continued management. Losses from neglect or purposeful infilling are great. In Britain, losses over the last 100 years from neglect or active infill have been severe, especially in areas of intensive agriculture, with over 60 per cent of all ponds lost in some areas, such as East Anglia (Table 2.1). At the same time, ponds have increased in urban areas, especially

Table 2.1 Estimates of pond losses from different parts of Britain over the last 100–150 years. Note that losses are particularly bad in areas of intensive arable agriculture. These figures refer to ponds that have been infilled. Those that remain may be so polluted or otherwise degraded as to be of little value to wildlife. The losses in urban areas are slight and there may be an actual gain in pond sites with many private garden ponds not identified during ecological studies.

Area of Britain	% lost	General land use
Bedfordshire	82	Intensive arable
Cambridgeshire	68	Intensive arable
Leicestershire	60	Intensive arable
Durham	41	Arable and pasture
Clwyd (Wales)	32	Arable and pasture
Midlothian (Scotland)	23	Arable and pasture
Edinburgh (Scotland)	6	Urban

small ponds in back gardens. In some areas urban ponds may now sustain the majority of amphibians and provide valuable habitats.

Many ponds are seasonal. Williams (1987) suggests the best classification for temporary waters is the length of the dry period as this determines their biota. The wildlife will have to show special adaptations to survive or recolonize. Tiny artificial ponds such as in tin-cans or water butts are referred to as containers, the equivalent of the tiny pools of water trapped by plant structures, called phytotelmata. Both have attracted interest as the breeding sites of pests, especially mosquitoes (Diptera), that transmit diseases.

Canals often have a similar wildlife to lakes and ponds. Although they possess linear features, their slow, interrupted flow and lack of erosive regime, coupled with disuse and shallow form, make most of them elongated ponds. Other largely artificial linear features such as ditches, leats, lodes and drains also tend to support pond biota.

2.3 Wetlands

The final stages in a lake's life, as vegetation smothers open water and terrestrial plants encroach, create wetland. These habitats are very important in their own right, though their ecology is not so well explored. In many parts of the world sprawling tracts of seasonally flooded land are only briefly open water, and essentially wetland.

Wetlands come in as many forms as open water. The primary classification depends on whether the wetland is on a mineral-based soil or peat. Mineral soil wetlands can be divided further between those in which the surface is typically inundated even during the growing season, which are swamps, and

Plate 7 View of Strumpshaw New Broad, a naturally nutrient-rich (eutrophic) water. Pollution, especially from sewage and agriculture, caused excessive nutrient levels and degradation in the present century (see Chapter 7). (Photo: Michael Jeffries)

those in which the soil remains waterlogged but are not flooded during the summer, called marshes. In Britain, swamps are dominated by tall emergent monocotyledons such as the common reed, *Phragmites communis* (Trin.), reed grass, *Glyceria maxima* (Hartman) Holmberg or great reedmace, *Typha latifolia* (L.), various sedges, *Carex* species and diversity of vegetation can be poor in these thick stands that smother lower growing plants. Marsh vegetation tends to be more diverse with many herbaceous plants as well as the tall reeds and sedges.

Peat wetlands form where decay of the dead vegetation is too slow to prevent considerable build-up of organic detritus. In many areas climate, geology and vegetation conspire to make the sites very acid, dominated by *Sphagnum* mosses and these are called bogs (or mires). They can be quite small features in a valley or basin or cover huge tracts of land as blanket bog. In some areas of lowland England peat also builds up but alkaline geology buffers against acidity and what appears to be a swamp community develops. These non-acidic peatlands are called fens.

A second category of wetland are those created by humans, which can be very extensive such as the huge tracts of rice paddies in the tropics. In Britain such habitats tend to be old agriculture systems, no longer used as they once were, nor maintained. The low intensity agriculture they supported was compatible with a diverse fauna and flora, so neglect or redevelopment of these sites is a conservation problem. They include flood meadows, relying on regular inundation to fertilize the soil and are used as summer pasture; washes, which were primarily areas to hold winter flood waters and keep the deluge off prime land; water meadows, which were similar to flood meadows but with a much more intricate and carefully controlled programme of irrigation.

2.4 Adaptation to the lentic environment

The ecology of animals and plants can be studied on many scales: individuals, populations, communities. The main theme of this book is to explore the processses that sculpt and develop whole communities and what happens when human activity intervenes.

Such factors generally operate on wildlife already well adapted to their environment. The basic problems of living in water have been overcome. Looking at the animals and plants as individual species is one scale and shows how they are adapted to survive in their environment. The adaptation of shape, physiology and behaviour at the level of the species is called their autecology, as against synecology, studying groups of organisms in relation to their environment and to each other. The physical and chemical environments of lakes and ponds described in this chapter present problems which species must have solved at the autecological level even before the wider interactions interfere. Aquatic organisms are adapted to the aquatic environment. Their autecology has been developed on an evolutionary time scale. There are some groups where major limits on their distribution can be credited to major autecological problems. For example, the paucity of coelenterates in freshwaters compared to marine habitats probably reflects the limits of their primitive physiology, though this raises the problem of how the few representatives such as *Hydra* (Coelenterata) species cope so successfully. However, they do not all live in all the waters. Though most groups of freshwater animals and plants are well adapted at the evolutionary, autecological level to survive the rigours and exploit the opportunities, their distribution and the precise composition of communities change, with the abiotic and biotic world around them. It is useful to consider some of the evolutionary adaptations at this stage, side by side with the environmental limits and opportunities that have moulded them, to see how beautifully and elaborately they are adapted, before exploring their synecology to see what factors actually do determine species' occurrence over shorter time scales.

Adaptations to lentic habitats are primarily solutions to the problems of water's odd character and to exploit the unusual benefits (Fig. 2.6):

1. Staying up in the water column rather than sinking into the hostile profundal. Staying in the euphotic zone brings added problems, for example avoiding herbivores and predators in such an exposed habitat.
2. Respiration, including survival in the sometimes anoxic depths.

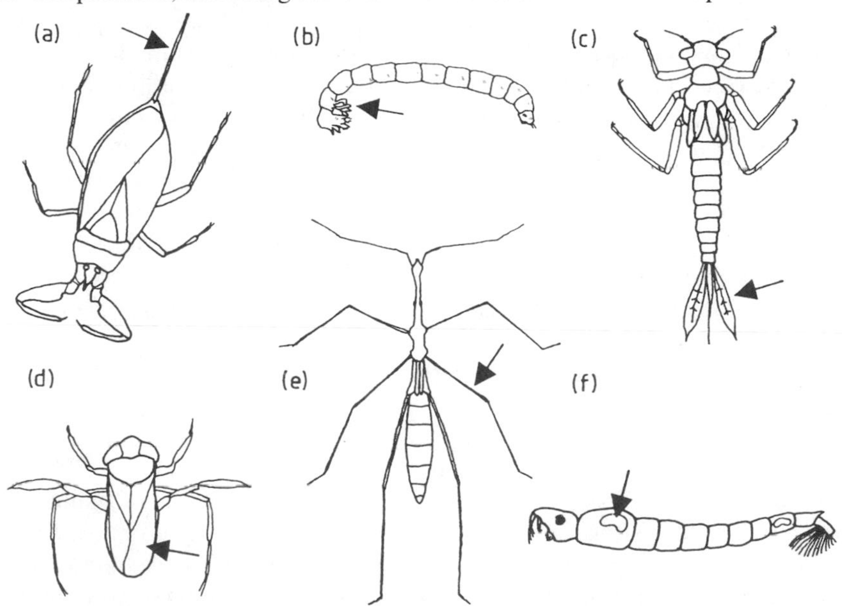

Fig. 2.6 Adaptations to the lentic environment. Respiration: (a) The water scorpion, *Nepa cinerea*. Hind snorkel tube to reach surface air. (b) *Chironomus* midge larvae. Live in oxygen-poor waters, with haemoglobin to maximize oxygen uptake. Tubule gills on rear segments. (c) Damselfly larva, *Zygoptera*. Tails are tracheal gills, with tracheal tubes enclosed in long plates providing large surface area to extract oxygen from water. (d) Waterboatman, Corixidae. Many such bugs and beetles use surface air trapped under wings and body hairs. Oxygen is extracted from the bubble and in part replenished by oxygen from the water.

Surface: (e) A water measurer, Hydrometridae. Very long legs spread this bug's weight.

Open water column: (f) Phantom midge larva, Chaoboridae. Hangs motionless in open water using gas vacuoles to adjust buoyancy. Semi-transparent to reduce threat from visual predators.

3. Survival in temporary habitats, at least when the habitat has vanished.
4. Exploiting the surface.

STAYING IN THE WATER COLUMN

Sinking out of the euphotic zone into the dingy profundal must be avoided by many creatures. Larger nekton can swim easily, though many fish are aided by buoyancy from adjustable gas-filled swim bladders. Many smaller animals and plants can swin. Even tiny plankton perform migrations to a diurnal rhythm and respond to cues such as temperature gradients. Other forms cannot move purposefully in open water. Pneumatic buoyancy is found in other creatures, using hydrostatic vacuoles of varying complexity. The phantom midges' (Chaoboridae:Diptera) larvae have adjustable gas vacuoles called pneumatostomes and can rise or sink with no swimming. Some protozoa (e.g. *Difflugia*) and algae (the blue–green algae) also contain vacuoles apparently used primarily for buoyancy. Many insects and molluscs which breathe by carrying a bubble trapped in water-repellent hairs, use this to surface simply by stopping swimming and letting go of the substrate.

Reducing sinking rate by size and shape is a passive adaptation. The shape of many phytoplankters (the colonial forms), development of spines and glutinous sheaths all reduce sinking rate. They all rely on turbulence and upwelling current to some extent. Many zooplankters also have spines and bristles. Such structures also interfere with herbivore and predator enemies.

Submerged plants benefit from the buoyancy of air-filled lacunae in their stems. These help with gas exchange, the buoyancy may be an added bonus.

RESPIRATION

Respiration has inspired many cunning adaptations. The small plankton, each cell or animal with a large surface area relative to body volume, are able to exchange gases freely without added devices. Some larger animals still rely on such diffusion, notably many worms, e.g. flatworms (Turbellaria), earthworms (Oligochaeta) and leeches (Hirudinea). These are either long and thin or long and flat. No part of the animal is far from the surface. Many are also tolerant of low oxygen levels. Oligochaetes are often pinkish in colour due to haemoglobin production.

Use of atmospheric air is still common, especially in the insects. They rely on a system of tubes, open to the air at spiracles on their bodies then ramifying as increasingly narrow tubes through the body eventually down to cellular level, for gas exchange. Many freshwater species have lost the open spiracles except for those on special snorkels which can be opened at the surface, for example water scorpion, *Nepa cinerea* L., and water stick insect, *Ranatra linearis* (L.), water bugs (Hemiptera), mosquitoes, rat-tailed maggots and soldier flies (Diptera). Adults of other insects, notably beetles

(Coleoptera) and bugs (Hemiptera) visit the surface regularly and take on board a bubble, trapped among water-repellent hairs and under wing cases. They can stay at the surface but equally dive or crawl down with the bubble, removing oxygen from it which is partly replaced by diffusion from the surrounding water. This system is called plastron respiration. The Hydrophilidae beetles have a more curious arrangement with antennae adapted to form a hydrofuge channel along which air is exchanged. All these air bubble systems serve spiracles open under the hairs and wings. Many pulmonate molluscs (Mollusca) can take air directly at the surface into a lung. Dragonfly larvae (Odonata) pump water into a chamber at the tip of their abdomen. The only genuinely submergent spider, *Argyroneta aquatica* (Clerck) (Araneae) relies on a body-hugging bubble caught in the dense hairs. This can be renewed at the surface but an air bell trapped in web acts as a replenishment source as well as refuge.

Pumping water in and out of narrow tubes and chambers is difficult. Lungs do not function well under water. Gills are their aquatic analogue. Gills work by passing water over thin walled plates or filaments projecting into the water, sometimes in a protecting gill chamber. Gills are best known in fish but occur widely in invertebrates: for example insects, crustaceans, molluscs and oligochaetes. Insects show particular variety. Those with gills have lost all open spiracles. The trachea tubes run underneath the thin gill walls and gases diffuse across. Mayflies (Ephemeroptera) have plates and feathers, damselflies (Odonata) flat tails, stoneflies (Plecoptera) tufts tucked under their legs, filaments in caddis flies (Trichoptera) and alderflies (Megaloptera) and tubules and lobes on many flies. All these variations enlarge the surface area for exchange, with tracheae lying just underneath the surface. Many species show gill-beating or body undulations to improve flow of water around the gills, varying this with oxygen levels in the water. The cases of caddis and tubes of midges may also improve circulation.

The animals of the profundal have to be well adapted to withstand low oxygen levels. Oligochaete worms are tolerant and often have haemoglobin. The larvae of *Chironomus* midges (Chironomidae:Diptera) are vivid red (called bloodworms) with pigment to maximize oxygen gleaning and undulate in their tubes. They can survive periods of anoxia. These creatures are associated with pollution, tolerating the reduced oxygen.

Excessive oxygen levels can be dangerous and occasional pollution deaths of fish have been ascribed to raised levels (Langford, 1983). This phenomenon is irrelevant to most life but the microbial biota of the profundal includes many anaerobically respiring forms. Some protozoa show migrations in the water column in response to oxygen changes, preferring poorly oxygenated areas.

Life at the water's surface is an additional habit, open to exploitation but requiring some adaptation. The general principles are to spread weight out and be water repellent to prevent wetting and sinking. Pond skaters (Gerridae:Hemiptera) and measurers (Hydrometridae:Hemiptera) have widely splayed legs and water-repellent hairs lending a silvery sheen to their undersides. Flies (especially the Dolicopodidae) and several spiders regularly hunt across the water, some of the latter supposedly using rafts. One beetle, the camphor beetle (a member of the essentially terrestrial Staphylinidae), has an extraordinary ability to skim across the surface, relying on a mechanism to delight any Victorian inventor. This is achieved by releasing water-repellent camphor from a gland on its abdomen. As the secretion spreads out on the water the beetle is pushed along. The whirligig beetles (Gyrinidae) are properly aquatic, but adapted to the surface. Their pair of compound eyes are divided so half of each one looks down below the surface, half above. The beetles twist in glittering swarms at the surface, but can dive below to safety. Many protozoa live underneath the surface film but their ecology is little known. A host of invertebrates live at the surface, often amphibiously, benefiting from the productivity. Molluscs, *Hydra* and small crustaceans use the underside to move across, perhaps grazing the underside. Beetles and fly larvae commonly dwell among emergent vegetation.

Plants of the surface are either free-floating, such as water fern (*Azolla filiculoides.* Lam.), duckweed (*Lemna* spp.), water hyacinth (*Eichhornia crassipes*), or the leaves of stems rooted to the substrate. The free-floating plants often have water-repellent surfaces and structures like hairs (in the case of *Azolla* extraordinarily divided trichomes) that resist wetting. Leaves rooted to the substrate are vulnerable to marked water changes. The growth form of most keeps the leaf flat on the surface and some, such as *Potamogeton natans* (L.) have a distinct hinge. These exposed leaves need surfaces able to resist exposure and are markedly toughened on the upperside. Many aquatic plants face temporary exposure of part of their structure and display different growth forms, submerged and emergent, even on the same stem. The precise variations differ. Some plants show a temporary change, toughening up essentially similar leaves to the submerged forms, for example starworts, (*Callitriche*), where exposed leaves are thicker and stouter. Others show much more discretely separate forms, for example water crowfoots *Ranunculus,* where submerged or emergent leaves are hugely different, long and feathery, submerged or flat and polygonal at the surface, or mares tail, *Hippuris vulgaris* L., where submerged leaves form gangling, plumed whorls, but, on the same stem above water, shorter, thicker spikes. These are a form of heterophylly (different leaved), a plasticity aquatic plants show in many circumstances where conditions change.

The final major physical problem in the lentic world is of dispersal, colonization and survival from year to year in temporary habitats. The dynamics, patterns and consequences will be dealt with in later chapters but the physical adaptations, especially for dormancy and resistance, are often remarkable in their own right. If ponds dry up many species become dormant in the mud. This dormant phase may be especially resistant eggs, for example Turbellaria, Oligocheata, many Crustacea, Ephemeroptera and Diptera. Resistant larvae, pupae and adults bury into the mud, for example Mollusca, Hirudinae, Hydracarina (mites), Odonata, Trichoptera, Coleoptera and Diptera. Where the buried form is simply the normally active stage in a state of lowered activity, not unlike the more familiar hibernation, the resting is termed aestivation. Other creatures enter a very different form to survive the inimical conditions. The most extreme example is the tun of tardigrades (water bears, Tardigrada). These are microscopic metazoans related to nothing else very closely. As conditions deteriorate the animal detaches tissues from the old body wall, which makes an outer layer and cells form a ball. When water returns, it rehydrates and re-forms (Williams, 1987). All these resistant egg cysts are also ideal for dispersal by the elements or hitching a ride on larger beasts. For many lentic plants resisting changes relies on the survival of roots and rhizomes which then sprout when conditions improve. Heterophylly, change to a clumped, toughened habit, is frequent among fringing herbs. Many deeper water plants are very intolerant of exposure and killed quickly. A greater problem faces the surface plants. In winter, icing on standing waters is common and the unattached plants would be very vulnerable. Some show adaptations sinking in winter and surfacing for summer, for example water soldier. Others form overwintering buds that are released in the autumn and form new shoots when conditions improve for example, bladderworts, *Utricularia* species.

3 Flowing Water

Rivers, streams and their like are called lotic systems. Their physical form, histories and life are dominated by a strong, unidirectional flow. The moving water coupled with their linear pattern creates other differences to the lentic world. Rivers rarely stratify. Instead, erosion and deposition are dominant processes. The flow, erosion deposition and resulting substrate are all closely linked and greatly affect the wildlife, from individual species' morphology and behaviour to the gross patterns of communities. Active river channels run downhill but the slope varies greatly along a river's course, especially longer rivers. Erosion lengthens the channel over time and the form changes along the length. The resulting landscape and habitats are collectively the river corridor and these are very responsive to the land and life of the surrounding catchment. Changes, natural or artificial, to channel, corridor or catchment can all alter the wildlife. As a river's course changes downstream so the habitat and communities alter. The effects of alterations and perturbations upstream can be carried into the downstream stretches far from the original source of the problem, inflicting the consequences at a distance.

So the ecology of lotic habitats differs from lentic systems by the dominance of linked flow, erosion deposition and substrate processes. Marked changes occur in the environment along a linear habitat which is sensitive to catchment changes. These can be rapidly transmitted to the river channel and further downstream.

3.1 The nature of lotic habitats

Just as the character and terminology of lentic waters reflect the dominant features of light, heat and resulting water masses so lotic habitats are best described by flow, erosion deposition and channel form.

Rivers and the adjacent land form distinct corridors in the landscape. This is particularly obvious in rivers with extensive flood plains forming a river valley. Sediment deposited during floods creates the plain, this typically flat terrain being alluvial. The actual river's course is its channel. In many corridors on shallow gradients rivers shift and migrate wildly and new channels are carved out. Old, abandoned channels, perhaps forming ponds or entirely dried-out, remain. From source to finish the channel has a longitudinal profile, with changing gradients generally steeper at source and increasingly shallow, though there can be marked variations at any point. Gradient is important in determining the streampower, the energy for overcoming friction and transporting sediment. The strip of land immediately alongside the channel is the riparian zone. The term is often applied to the fringing vegetation, for example riparian vegetation, but can be used more generally, for example riparian landowner (Fig. 3.1).

Channel form varies with time and place along the river, equivalent to the successional changes in lentic habitats. River channels are dynamic features. They react to landscape and topography, for example gradient changes across geological strata or sediment of different resistance to erosion, to reach a general equilibrium, a speed and size, that continues for many years. Within this overall regime individual microhabitats and channel features come and go with the years. The channel has a dynamic equilibrium. Major changes in the equilibria can occur, most obviously in response to gross geological change. In northern temperate areas many rivers seem too small for the size of the valleys they occupy. The landscape reflects the volume of water as the glaciers retreated and the equilibrium as it existed then. Other channels seem suddenly steep and deep, reflecting rises in the land mass and a resulting increase in erosion by the stream towards a new equilibrium. A

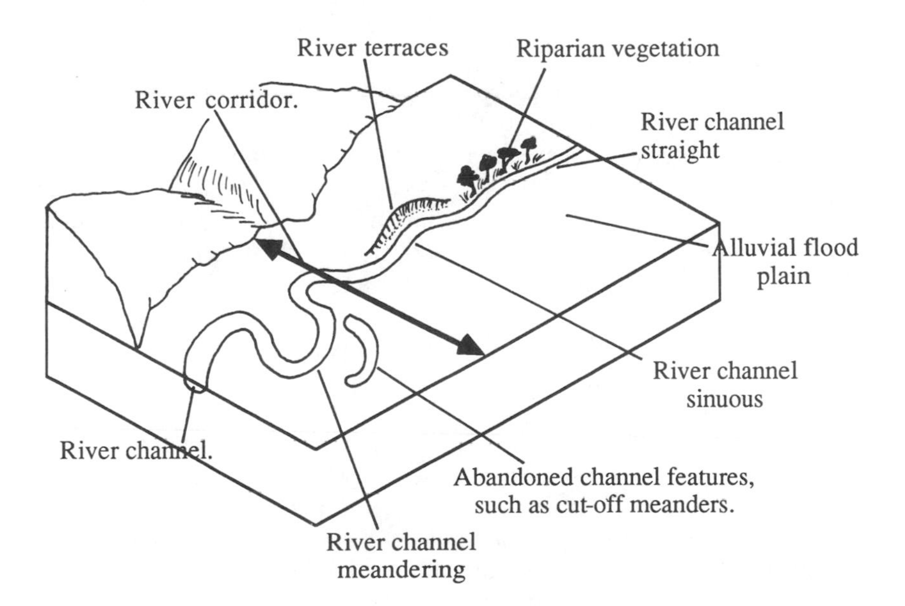

Fig. 3.1 Gross features of a river corridor

more recent threat to channel equilibrium is from impoundment and regulation of stream flow altering the downstream regime. Craig and Kemper (1987) review the ecology of regulated streams.

River channels contain a mosaic of habitats, largely created by flow-substrate interactions. Many channels are sinuous, eroding into the outer banks of bends, depositing on the inner banks, active meandering or a more passive response to confining points of resistant geology, confined meanders. Less sinuous channels may still modify with braiding, the formation of mid-channel bars exposed as islands at low flows, with a network of channels in between. Inactive stretches, apparently unchanged in historical times, are not uncommon in some areas. The flow– erosion–deposition regime sculpts the channel bed too. The typical pattern is of alternating shallows, with coarse substrate and often turbulent flows, called riffles, and deeper, despositing areas of smoother, laminar flow, called pools. These may form as flow alters in response to channel direction, pools being gouged at the eroding outsides of meanders, riffles along the sheltered inner bend. They also occur in straight channels and a distinctly regular pattern is typical with riffles (or pools) spaced at intervals of 4–7 times the channel width (Fig. 3.2). Quite why is unclear. Once established they are self-sustaining, as long as the gross equilibrium is not radically changed. Erosion and deposition can be triggered by small events such as a fallen tree or a single boulder. Once the rhythm of riffles and pools is created it remains until another disturbance. Riffles and pools created at unnatural intervals, in defiance of the natural regime, are soon lost. Both pools and riffles are distinct habitats with differing communities and are commonly distinguished in practical studies. Other gross habitats have been described and harbour their particular life. Deep riffles, lacking the turbulent flow, are called runs. Sheltered bays along the banks are slacks and backwaters are channels lacking the unidirectional flow, certainly for most of the year. Vegetation adds diversity either as aquatic macrophytes or tree roots. In Chapter 6 an example of animal community associations with these habitats is given. Channel edges have a distinct fauna compared to mid-channel (Ormerod, 1988). Individual boulders or fallen trees create obstructions and provide different microhabitats. All these fairly simple criteria, difficult to define in precise, measured terms but obvious to the trained eye, are increasingly important in the practical conservation and management of rivers (Lewis and Williams, 1984; Nature Conservancy Council, 1985).

The flow of water encompasses several different ideas. The volume flowing through the channel is the discharge, typically measured as cubic metres per second. The actual speed at which it moves is the velocity. The two are linked by the cross-sectional area of the channel. In a wide river, discharge can be high, though velocity is low, the river appearing calm and sluggish because the channel's width allows so much water to travel. In a narrower channel the same discharge can only be achieved if the velocity is much faster, allowing the water to move quickly through the smaller

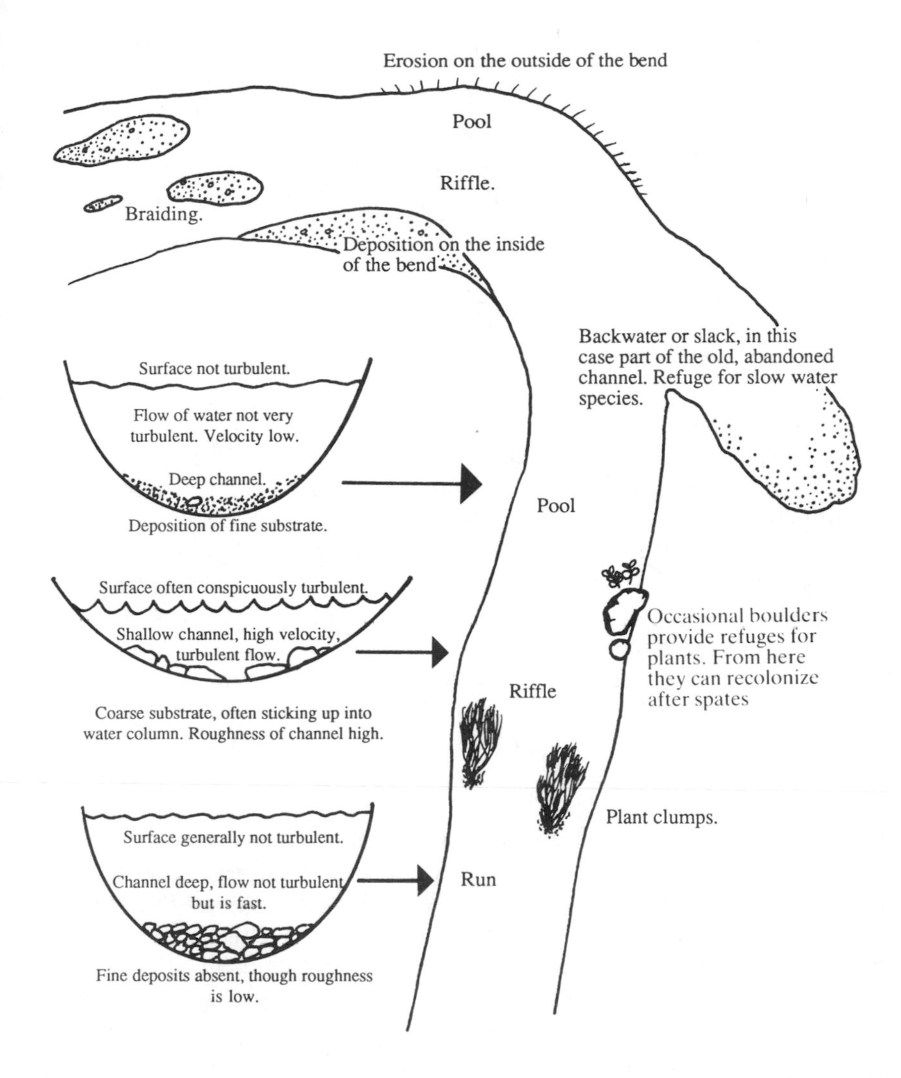

Fig. 3.2 Habitat features of a stream channel, with cross-sections of run, riffle and pool sections.

channel. At its simplest, discharge (D) is a function of velocity (V) and cross-section (A), $D = AV$. Discharge is often called streamflow. The velocity of the current is not the same across the channel. Friction slows the water down over the channel bed, along the sides and under surface ice. The fastest velocity is found in the centre of the cross-section. The reduction of velocity is so great immediately above the bed that the 1–3 millimetres directly over and around the substrate is a boundary layer. Similar reductions occur in the lee of obstructions. The boundary and shelters provide a microhabitat out of the worst of the current. The form of the channel bed, substrate, debris and clumps of plants all contribute to the channel's roughness. Unaltered rivers are often very rough, essentially possessing many and varied habitats. Reduction of roughness by river engineering to increase discharge for drainage and flood prevention is a major cause of degradation.

Discharge fluctuates especially to a seasonal rhythm. The changes alter flow, deposition and erosion, perhaps the gross channel form. The discharge echoes the precipitation into the catchment, the water in the channel is part of the basin hydrological cycle (Beaumont, 1975). Essentially, discharge equals the precipitation minus evaporation loss, groundwater retention and soil moisture. Rivers differ in the range of discharge variation and timing of high and low flows. The pattern of discharge is called the river's regime. Discharges that fill the channel without flooding over are bankfull. Inundation of the corridor occurs at higher discharges and the extent of floods is often described by their periodicity. Many channels have minor flooding each year. Larger deluges occur at increasing intervals, ten-year, fifty-year, one-hundred year floods. Periods of high and bankfull discharge are most important for erosion. The seasonally flooded land represents important habitat. Extensive tracts of tropical rain forests are seasonally flooded providing rich feeding and breeding grounds for fish (Lowe-McConnell, 1975; Williams, 1987). The lingering pools create a multitude of ephemeral lentic habitats. Even in intensively drained and managed landscapes controlled flooding maintains unusual systems such as watermeadows or paddies.

The debris moved by the river, its load, plus suspended algae make the water turbid. Changes can be very sudden if detritus is washed in from the catchment and the effect will move downstream, interfering with stretches far from the source. The force of water needed to move different sized particles varies, so load and deposition alter with discharge. The river sorts the particles by size. The precise composition of the sediment and the rate at which it accumulates greatly affect the wildlife. The mixture (heterogeneity), stability, texture and organic content are all separate factors (Minshall, 1984). Size composition can be measured very precisely with the Wentworth Classification of Substratum Particle Size (Hynes, 1970; Minshall, 1984). The actual sizes are transformed into $\log_2$, which allows the wide range from silt to boulders to be converted into an arithmetic series (Table 3.1). This is important in studies of animal and plant ecology but often impossibly time consuming for much applied work. Intuitively, obvious classification schemes have been developed and used successfully to grade substrate by eye, into categories such as bedrock, boulder, cobble, pebble, gravel, sand, clay (Lewis and Williams, 1984; Nature Conservancy Council, 1987). The animals on the substrate are the benthos, just as in lentic waters. The zone down in the substrate has been called the hyperheic (Williams, 1984). A surface film of algae, fungi and bacteria coats the rocks, macrophytes and other detritus, and comprises the periphyton. Plankton

Table 3.1 Substrate particle size classification. The diameter size ranges are given along with Phi scale category. The Phi scale is $-\log_2$. Particle size categories are often gauged subjectively for conservation work and the equivalent terms for the measured categories are also given. Sub-divisions such as coarse or fine are seldom used in these practical studies

Particle size diameter (mm)	Phi scale category	Category name
--	--	Bedrock
>256	−10, −9, −8	Boulder
128–256	−7	Large cobble
64–128	−6	Small cobble
32–64	−5	Large pebble
16–32	−4	Small pebble
8–16	−3	Coarse gravel
4–8	−2	Medium gravel
2–4	−1	Fine gravel
		Sand:
0.5–1	0	very coarse
0.25–0.5	1	coarse
0.125–0.25	2	medium
0.063–0.125	3	fine
0.032–0.063	4	very fine
<0.032	5, 6, 7, 8	Silt
	9,10	Clay

are less characteristic of lotic systems. Plankton do occur in slow flowing reaches and sheltered backwaters. From such areas they are washed into the main channel. Dislodged periphyton and lentic plankton washed from upstream lakes add to the soup in the water column. However, these populations are scarcely living in the habitat but rather being carried through it and riverine plankton (potamoplankton) only truly thrive in sluggish waters.

Although stratification is unusual in rivers, given their turbulence, movement and relative shallowness thermal conditions do vary. Seasonal and diurnal changes occur, and thermal regimes vary along the length of a river with topography and volume. Marked differences can occur where confluence channels bring together water masses of different temperatures or loads of suspended solids that will not mix easily. Famous examples occur in the Amazon system where so-called black-water and white-water rivers join and the lack of mixing is very obvious due to the different coloured water masses. Spring-fed streams show a characteristic regime. Hot (thermal) springs cause viscously hostile conditions, for example temperatures up to boiling point (100 °C), and detectable concentrations of toxins such as arsenic, recorded at over 1.5 mg l^{-1} in some instances, immediately downstream. These extremes are inimical to all life at first, then a zonation of increasingly tolerant organisms develops as the water cools and many minerals crystallize out (Castenholz and Wickstrom, 1975). The majority of springs supply water from underground largely protected from seasonal changes in temperature. Hence spring-fed streams tend to be cooler in summer and warmer in winter than expected. Light penetration varies with turbidity and a little with surface turbulence but generally light is not a dominant influence. Shading from the banks and riparian vegetation can be again the narrow linear form of lotic habitats making them susceptible. Glacial rivers often carry very heavy loads of suspended solids and light hardly penetrates.

Gaseous exchange, unhindered by stratification and positively encouraged by turbulence, mixing and cascades, generally maintains an equilibrium. The result is that many lotic animals are adapted to high oxygen levels, or, rather, have no adaptations to cope with low levels, making them vulnerable to changes that would not harm most lentic wildlife. Oxygen varies with photosynthesis, with production in the day and a lag at night, pulsing downstream. Oxygen in the substrate may be a problem, especially if compositional changes clog the pores preventing interstitial currents.

Other nutrients, dissolved ions and pH influence rivers and streams much as they do lakes and ponds. Differences arise more from the greater and faster input into rivers from their catchments than into lakes, at least in the short term. Discharge changes alter dilution and catchment run-off may shift the balance of chemistry. Chemistry changes from source to finish and impacts downstream. This applies to organic input with allochthonous sources of prime importance in many streams. The progressive use of organic debris downstream is another source of variation along a course, as particle size and decomposition alter.

There are many good texts describing the hydrology of rivers and their landscapes, for example Crickmay (1974), Whitton (1975), Lewin (1981), Richards (1982) and Morisawa (1985).

Plate 8 The large glacial river Thjorsa in south Iceland. The photo depicts the braided river channel with large unstable banks of glacial deposits. (Photo: Derek Mills)

3.1.1. *The biography of rivers*

Rivers change with time, their habitats altering. The rhythms and threads of time are recorded in the channels' changes and their linear form results in different stretches of the river being at different stages of development. Rivers evolve with time, changes occurring on different scales from hours and days to geological epochs. Studies of whole catchments are largely palaeohydrology. Changes of specific channel stretches, the metamorphosis of a river, can be done by direct observation, from old maps and recent landforms. Some changes, particularly of the channel, arise from the river's regime itself, so are autogenic. The effects of climate, catchment and perhaps human activity are allogenic. As with lakes, some large rivers are very old: the Nile has flowed since at least the early Eocene, the Congo since the Pliocene (Beadle, 1974). Even rivers in nothern latitudes, that have only existed in their present form since the glacial retreat, may flow in drainage patterns that are much older, for instance Scotland's drainage pattern may have its origin in Cretaceous times (Sissons, 1976).

Just as the biography of a lake, from barren youth to verdant old age, is something of an ecological just-so story, full of variation in reality, so lotic systems have a general history. The overall gradient pattern is from steep to shallow. Headstreams are generally straighter, meanders and migration increasing downstream and with age. At any stage in a channel abrupt changes in gradient will cause exceptions. Many steep, montane streams suddenly falling to highland valley floors meander widely. The final stretches of other streams, notably short systems coming off steep terrain, remain straight down to sea level. Lowland rivers of limited streampower and on cohesive substrate also remain straight. The division into youthful headstream, middle and lower courses provides a useful synopsis of river history and changes.

Steep headstreams, either of youthful rivers or feeding older channels in lowlands far away are steep and unstable. The current is apparently fast, though high roughness and limited volume limit streampower. There is little fine sediment from the steep valley. Changes in discharge can be rapid and these streams tend to be spatey. Few macrophytes can survive. The turbulent, cool water is well oxygenated. Catchment chemistry can have rapid and overwhelming impact. The small channel size means individual boulders, fallen trees and bank collapses have a big effect on channel and habitat. These streams are unstable and eventually some sinuosity will be initiated. The development of this widens the valley into a corridor.

The middle and lower reaches are harder to distinguish. Middle reaches are dominated by transfer of material, the lower reaches by deposition. Both have decreasing gradients, greater discharge and bigger channels. Their form is marked by increasing sinuosity and activity (the variations in erosion and deposition). Different types of sinuosity occur. Active meanders, confined meanders, low sinuosity and inactive channels. The flow is less turbulent in the deeper channel and concentrations of suspended solids higher. The build-up of nutrients from catchment inputs is greater. Shallower gradients allow fine particle deposition, again a nutrient source. The actual velocity may still be as fast, or faster, than the apparently rapid headstream, but less turbulence, slower changes in discharge, substrate accretion and nutrient input all allow macrophyte growth. Different plants are typical of different regimes and their distribution changes over time, moving upstream as gradients erode away. The river corridor in the lower stretches becomes a higgledy-piggledy plain of reworked deposits, patches of riparian vegetation, abandoned channels and dead-end arms. Extensive wetland habitats commonly develop alongside and where the channel fans out at estuaries. The river has become the main creative force in the landscape, contrasting with the headstream vulnerable to even small-scale catchment inputs.

As rivers normally exist in a dynamic equilibrium with the landscape, changes to the whole river system happen only slowly. Headward erosion extends the drainage system at the headstreams and downstream stretches mature into larger channels. Rivers do not vanish in the way lakes and ponds do. Lotic habitats are long-lived, though the precise channel patterns alter as each stretch ages. Dramatic changes occur in response to major perturbations, rapid uplifting and the comings and goings of glaciers. Rivers may last for a very long time, though humans have shown a remarkable ability to obliterate them. The precise age of a river may not be particularly important. The rate at which the basin is denuded towards complete destruction and the state of the equilibrium between discharge, gradient and channel are probably more important for describing the condition of a river.

3.1.2 *River classification*

Many attempts have been made to classify types of rivers, both the whole drainage system and the different stretches from headstream to lower reaches. Schemes have been increasingly motivated by the need to describe rivers for conservation and management, to characterize the flora and fauna expected at a certain stage, if healthy, and quantify damage.

One of the oldest classifications divides stretches into youthful headstreams, mature middle reaches and old age, reflecting the physical channel forms described in the preceding section. The implication of energy available at each stage is deceptive. Apparently sluggish old age reaches have such a high volume that energy is great. An alternative to this three-fold division uses river load processes, sediment production, transfer and deposition.

A geomorphological description for whole systems that is widely used is stream order. The order is based on the hierarchy of how channels link up. First-order streams have no tributaries, they are the first streams fed by

Plate 9 The White Water, flowing off the eastern slopes of the Scottish Grampian Mountains. The fish in this reach of the river are predominantly salmon and trout. It can be seen that the lower slopes of the mountains are afforested. Low-lying mist is a source of occult deposition. Note lack of channel macrophytes. (Photo: Derek Mills)

springs, groundwater and run-off. When two first-order streams join, the confluent stream is second order. Two second-order streams joining make a third order, two third-order streams make a fourth, etc. Any stream joining a higher order channel (e.g. a first into a second, third into a fourth) does not result in an increment in stream order (Fig. 3.3). The ordering hierarchy is at least objective and simple, even if detail is hidden, and has proved a valid factor, in analysing community patterns.

The animals and plants themselves have been used to identify zonation. Responding to the physico-chemical world and having to survive all year round the wildlife has great potential for typing habitats or as indicators of change. Sampling of water quality itself is time consuming and it is impossible to measure all the variables. The animals and plants are doing the monitoring for you, they are there all the time, integrating responses to all factors.

Zonation of fish distributions has been used to provide a simple classification, akin to the youthful, mature and old-age physical zones. The typical pattern in temperate streams is (from headstream downwards) trout (*Salmo trutta* L.) zone, grayling (*Thymallus thymallus* L.) zone, barbel (*Barbus barbus* L.) zone, bream (*Abramis brama* L.) zone (Huet, 1954).

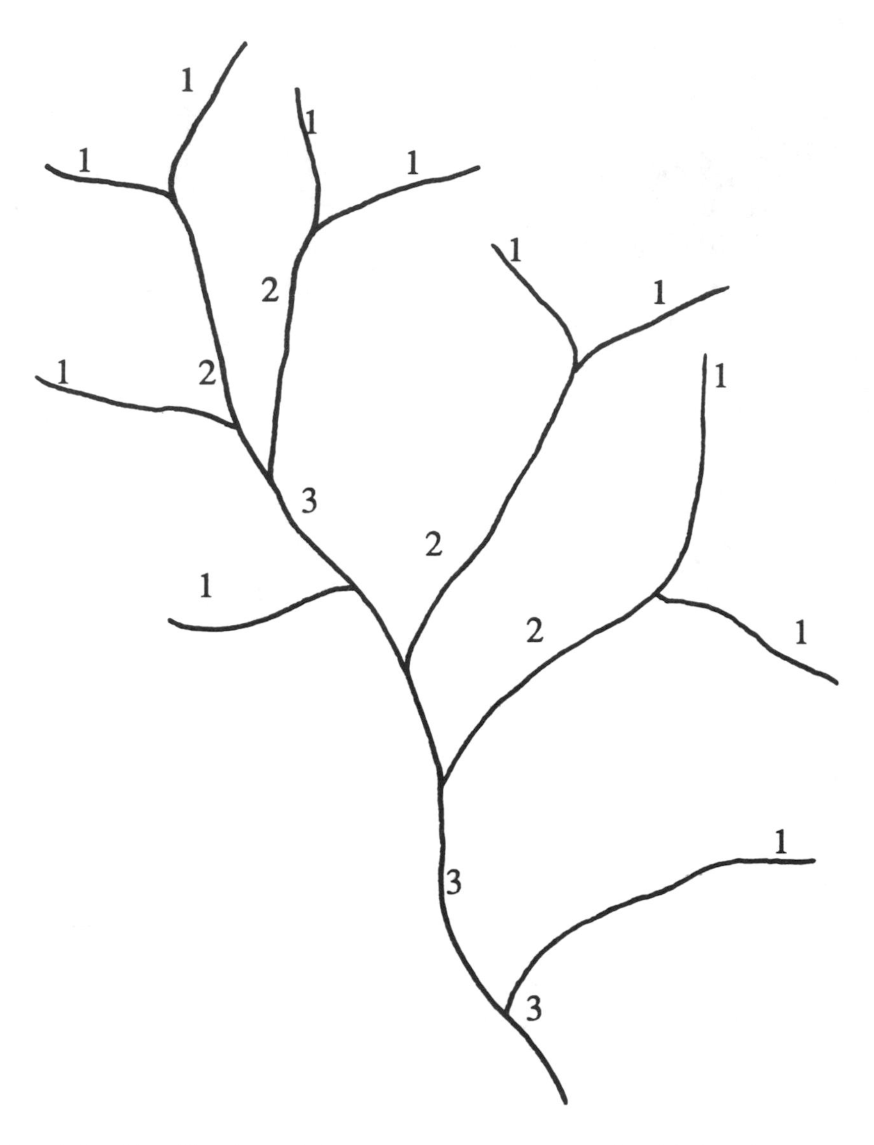

Fig. 3.3 Classification of river system by stream order. First-order streams are initial headstreams, with no tributaries supplying them. Where two first-order streams join the confluent channel is termed second-order. Where two second-order streams join the confluent channel is termed third-order. Note that a lower order stream joining a higher order stream does not affect the order rating.

Alternatively, headstream (no fish), trout zone (troutbeck), minnow (*Phoxinus phoxinus* L.) reach and lowland reach (Carpenter, 1928) have been suggested (Fig. 3.4). Geographical variation in fish distributions, even

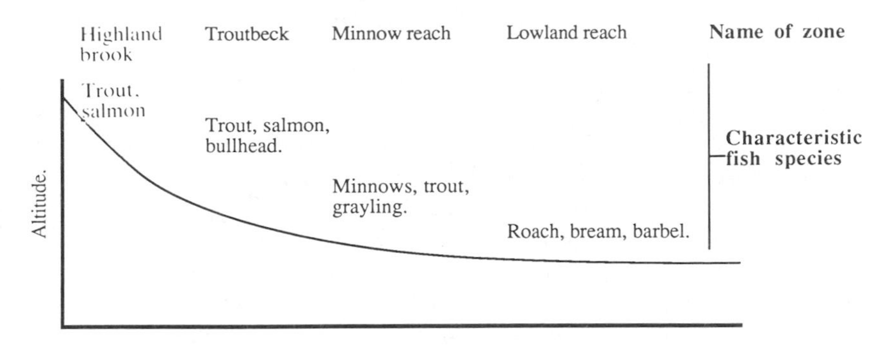

Fig. 3.4 Temperate river zone classification based on fish species. (After Carpenter, 1928)

in one country, render such schemes useful only on a local scale. Mills (1989b) found that this scheme was appropriate for the River Tweed which is a mature river possessing all zones. Nearly every group of stream invertebrates shows a zonation of species along channels and many invertebrate river typing schemes have been devised. An early attempt to look for universal patterns globally was an analysis of invertebrate changes along channel (Illies; the original papers are mostly in rather inaccessible European journals but are well summarized by Hynes, 1970). Illies identified zones separated by marked faunal changes, especially complete loss or appearance of taxa. One such break, generally equivalent to the downstream limit of the grayling zone, proved very consistent. From headstream to this divide streams were characterized by monthly mean temperatures less than 20 °C, high oxygen, fast and turbulent flow over coarse substrates. This stretch was called the rithron. Below the divide monthly mean temperatures often rose above 20 °C, oxygen was variable with occasional severe sags, smooth flow and fine substrate deposits occurred. This was called the potamon. Attempts to subdivide further the rithron and potamon have been made but the breaks are not distinct (Hynes, 1970). Sharp divides are difficult to specify, communities grade into one another. This gradation prompted development of the 'River Continuum Concept' (Vannote *et al.*, 1980). Communities were described by their position along the river and also trophic structure (the variety of feeding habits among the invertebrates and their abundance). The concept primarily describes the alteration in communities with energy transfer. In the headstream communities relying on the large debris, such as allochthonous leaves, thrive. Downstream communities dominated by animals utilizing fine particles, either allochthonous debris broken down upstream or autochthonous microflora production, take over. Nutrients are moved down the system through a sequence of communities, each one cycling some of the energy. The result has been called the Nutrient Spiralling Concept, closely linked to the River Continuum. The Continuum Hypothesis is still the subject of debate. Minshall *et al.* (1983b) put it to the test across several streams and found that the idea stood up but with complications from inter-regional differences. However, neat gradations are not always found.

Interest in classifying rivers to monitor pollution has generated a lot of schemes, generally forms of biological monitoring where the variety and abundance of species hopefully reflect water quality. There is a wealth of pollution indices. Invertebrates are the most widely used. Most indices combine the presence or absence of taxa, their abundance and some weighting of their importance, based on tolerance to pollution, to arrive at a numerical score. The score for a sample, or site, reflects the level of pollution. All types of wildlife, from viruses to mammals, have been suggested for biomonitoring (James and Evison, 1979; see Chapter 9). Ecological diversity measures, originally devised for general ecological studies, can also be used in this way as they score the variety and abundance of wildlife. Hellawell (1978, 1986) provides an extensive review of many of these schemes.

Sampling whole communities provides huge data sets: many species spread across many sites and pollution indices are a convenient way to simplify these results into a useful form, although some information is lost. Also such indices do not sensibly distinguish between the tremendous variety of clean water communities, interesting in their own right. In recent years great advances have been made with multivariate analysis techniques. These are analyses capable of condensing down data sets of many variables (hence multivariate) to pick out the important patterns. Not only do they adequately analyse many variables but these variables can be very different types of parameters (animal, plant, physico-chemical) measured in very different ways (abundances, biomass, chemical analyses) and the analyses can still look for patterns of similarities and differences. The techniques often help pick out indicator species that are consistent representatives of a particular community and use mathematical measures of similarity (or dissimilarity) to quantify the results. Most such analyses have only been widely developed with the advent of computers allowing rapid handling of the large data sets. There are two broad categories of analysis. Firstly, techniques that divide samples into discrete sets, either by starting with them all lumped together and progressively dividing them into smaller and smaller groups, based on their comparative similarities (or dissimilarities) or starting with all samples isolated and progressively linking them together in clusters. These are called classification techniques. The alternatives take all samples and arrange them in patterns, again reflecting the degree of similarity or difference. Such patterns can be spread along one, two, three or more axes, the coordinates being a relative measure of the similarities. These techniques are called ordinations. They can be used to measure all

Plate 10 The River Tyne, East Lothian, Scotland. The river rises from nutrient-poor upland geology but flows through a rich agricultural catchment. The result is an oligo–mesotrophic habitat rich in channel and emergent macrophytes (see Chapter 3). (Photo: Michael Jeffries)

Plate 11 The River Frome, a chalk stream in Dorset, southern England. Beds of *Ranunculus penicillatus* var. *calcareus* (Butcher) are much in evidence. The photo was taken prior to the annual practice of weed cutting undertaken on many southern chalk streams. (Photo: K. Morris)

sorts of ecological patterns from extensive studies comparing many different rivers (Wright *et al.*, 1984; Corkum, 1989) to distinguishing habitats in the same stretch of one channel (Jenkins *et al.*, 1984, see Chapter 5; Barmuta, 1989). In Britain a nationwide typing of clean rivers, using multivariate classification and ordination of invertebrates, has recently been completed (Wright *et al.*, 1984; Furse *et al.*, 1984). This data base has been tested by Water Authority biologists (now part of the National Rivers Authority), to refine the detail and allow for regional variation with the intention that it can be used for conservation and pollution monitoring.

Macrophytes can be used to type rivers, though less attention has been paid to them (see Table 3.2). Multivariate techniques have been applied in Britain to develop a nationwide river typing scheme. The plant communities reflect flow, substrate, trophic status and general catchment characteristics (Holmes, 1983; Holmes and Newbold, 1984). Haslam and Wolseley's (1981) scheme has been refined so that likely plant communities on an unsurveyed river can be predicted from geology, topography and channel width alone. Significant departures from the prediction in the real survey data can be used to spot and assess damage. With the increasing realization that rivers cannot be wantonly separated from the surrounding landscape for conservation and management, survey techniques for channel and surrounding land combined are now available (Nature Conservancy Council, 1987; see Fig. 3.5). Simple, wide-ranging and sensible descriptions of channel features,

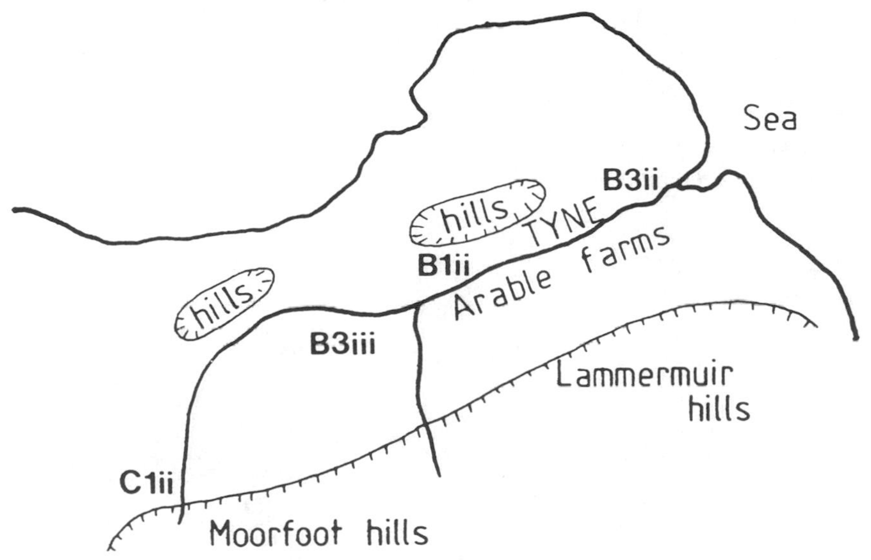

Fig 3.5 The River Tyne catchment in the Lothians, Scotland. The river rises in upland terrain, resistant geology. However the geology is not barren, the Lammermuir and Moorfoot hills are ancient sedimentary rocks. The uplands are largely poor grazing. Catchment land use in the main river valley adds to the river's trophic status with intensive arable and grazing. There are several small towns along the river. The upland but slightly enriched character is borne out by macrophyte and invertebrate classifications. The codes C1ii, B3iii, B1ii and B3ii are part of a river classification system based on macrophytes and derived from rivers throughout Britain. The general characteristics of the classification codes along the Tyne are described in Table 3.2. The River Tyne is shown in Plate 10.

Table 3.2 Classification of sites on River Tyne, Scotland using macrophytes (Holmes and Newbold, 1984) and invertebrates, (Wright *et al.*, 1984). The categories refer to identification codes in the original publications. General descriptions of site category characteristics, derived for rivers throughout Britain, are given. Compare to the Tyne's actual details given in Fig. 3.5.

R. Tyne sites from macrophyte classification.

Site category	C1ii	B3iii	B1ii	B3ii
General description of site type	Upstream sites of upland rivers. High velocity but base and nutrient rich.	Base rich, but velocity and flow characteristics maintain course, nutrient-poor substrate compared to lowland rivers	Deposition of finer substrate increases nutrients. Additions from agricultural sources	Fine sediment deposition
Typical macrophytes	*Phalaris arundinacea* *Mentha aquatica* *Agrostis stolonifera*	As C1ii plus *Rorripa nasturtium-aquaticum* *Veronica beccabunga* *Myosotis scorpiodes* *Sparganium erectum*	As B3iii plus *Callitriche stagnalis* *Elodea canadensis* More large algae, lichens, liverworts and mosses	As B1ii plus *Ranunculus fluitans*
Total number of macrophyte species	31	39	36	43

R. Tyne sites from macroinvertebrate classification

Site category	17	19	20
General description of site type	Upstream sites on upland rivers. Coarse substratum, little macrophyte cover, low alkalinity	Middle and lower reach sites of upland rivers	Middle and lower reach sites of upland rivers, but characterized by lower slope, less coarse substrate, higher macrophyte cover and higher alkalinity than true upland rivers
Number of regular, abundant species	35	30	29

corridor landscape and vegetation can now be made.

There are very many ways to describe and classify lotic habitats. The choice depends on the type of information needed and what use you hope to make of it.

3.2 The riparian habitat

Rivers often flow through and are responsible for wetlands similar to those surrounding and smothering lakes. Marshes, swamps, bog and fen swaddle the courses of many rivers, especially in low gradient reaches prone to flooding. River flooding regimes are vital in the creation of habitats other than the natural wetland. Wet grasslands such as flood meadows, washes and water meadows are all human-managed; their special ecology, history, beauty and continued survival are a result of that management. The simplest, in management terms, are areas set aside to hold flood waters that threaten surrounding land, often called washes. In winter excess water smothers the rough pasture and slowly drains away once danger is past. The meadows can be grazed in summer. Grazing meadows may flood but are primarily maintained for livestock all year, their dampness and dissecting waterways adding wildlife interest. Water meadows are complex systems with carefully controlled inundation (compared to the gross flooding of washes) used to maintain fertility for arable farming. Inundation of rice paddies, fish ponds and similarly productive land in the tropics sustains an equally complex heritage of land uses.

The riparian vegetation along the riverbanks is a unique habitat, existing in a mutual balance with the channel. Riparian vegetation is not simply riverside forest, the community is distinct (Mason *et al.*, 1984). The vegetation alters many features of the waterway ecology; abiotic factors such as light and temperature, bank stability, allochthonous input of plant and animals, autochthonous productivity (by shading), a physical buffer to run-off, especially as a sediment trap, and as physical habitats for birds and mammals, overhanging shelter for fish and somewhere for emergent insects to rest, feed and lay eggs (Table 3.3). Riparian vegetation is as vulnerable to careless management as the river channel itself. It is also prone to natural destruction and change. The riparian zone is often a kaleidoscope of reworked river deposits and different successional communities of plants (Kalliola and Puhakka, 1988). The heterogeneity of age and structure benefits the variety of wildlife. Physical changes to the river regime or channel morphology alter this mosaic. Engineering works have proved very damaging (Newbold *et al.*, 1983). Care is now taken and engineering activities can be designed to maintain, even create, habitat diversity (Lewis and Williams, 1984; Raven, 1986a, 1986b). Much of the practical conservation work along rivers consists of riparian management and restoration (Lewis and Williams, 1984).

Table 3.3 Impact of riparian habitat on stream ecology. Data from river with conifer plantation and meadow habitats along its banks. The data refer to summer. Note how restricted light affects moss cover, numbers of invertebrates and trout productivity. (After Smith, 1980)

Riparian vegetation	% illumination of stream surface	% moss cover of substrate	Macroinvertebrates		Trout	
			No. taxa at sample sites (from 5 samples)	No. individuals, total (from 5 samples)	Abundance (No. m^{-2})	Biomass (g m^{-2})
Conifer forest.	19	5–30	33–40	1000–1400	0.09	6.74
Meadow	89	46–57	37–44	3000–4000	2.06	12.3

3.3 Adaptations to lotic life

Many problems and answers are shared by lotic and lentic wildlife, for example, adaptations for respiration and osmoregulation.

The particular adaptations for living in fast-flowing rivers and streams are to resist or avoid the strong currents. Only fish, birds and mammals can swim far or effectively against the current. Most animals and plants actively avoid the open water. Physical adaptations are many-fold. Lotic animals commonly have flattened bodies and limbs. Mayflies of the family Ecdyonuridae are famed examples but dragonflies and caddis flies show this and other animals exploit their naturally flattened form, for example, stream flatworms (Turbellaria). This allows these creatures to live in the boundary layer around the substrate, often on the exposed upper surfaces. Streamlining occurs in those fish (e.g. trout) frequenting the faster water. A very similar body form is found in some mayflies (e.g. Baetidae), minimizing drag. Some caddis lash twigs along the lengths of their cases, projecting some way behind. These act as vanes, turning the animal into the current as it clings on with its front legs to present the least resistance to the flow. Suction devices are widespread across taxa. Some, such as leeches, normally have oral and anal suckers and can use this pre-adaptation (a more specific term for an adaptation, already fully functional for one purpose such as the leeches' suckers and that can be turned to exploiting another opportunity or solving a different problem is an exaptation; Vermeij, 1982). Snail and limpet species rely on the suction power of their muscular foot. Similar devices are friction pads and sculpted body edges that can be held flush with the substrate. Riffle beetle (Elminthidae) larvae, with sculpted exoskeleton plate rims, for example *Elmis aenea* (Müller), and many mayflies, with enlarged gills or bristle pads, utilize this adaptation.

A good grip from claws, hooks and bristles is common to many taxa. The curved claws on the anal prolegs of caseless caddis are a fine example. Cased caddis often have strong claws too but on much shorter legs, simply anchoring themselves inside their cases. The net-spinning caseless species not only use their silk spinning to catch food but provide a refuge from the current. The combination of silk and refuge is used by other insect larvae, chironomid larvae, moth caterpillars (Lepidoptera) and black-fly larvae (Diptera, family Simulidae). The black-flies are characteristically a lotic group. Larvae spin silken pads on to the substrate and anchor themselves to this with circlets of hooks on their behinds. The head can be held out into the current and antennae used to filter food. An emergency line is kept attached to the anchor point from the head silk glands. If swept away the larva reels itself in using jaws and stumpy prolegs to grip the thread. Species of many insect orders pupate underwater (Trichoptera, Diptera, Lepidoptera). The pupae must be firmly stuck down with glue or silk. This then causes a problem of how to emerge and reach the surface. Caddis pupae have remarkable mandibles that might be used to slice open the pupal case.

Ballast is used by molluscs with heavy shells and some caddis which add a few larger pebbles to otherwise normal cases. The riffle beetles normally clamber about on the substrate using plastron respiration to breathe. Occasionally they must refresh the bubble and can finely adjust their buoyancy to float up. They live in fast streams so must descend as soon as possible, again relying on bubble adjustment to do so (Fig. 3.6).

Behavioural adaptations help. Animals can be very sensitive to precise flow and avoid or select sites accordingly. Many live under rocks and stones or burrow deep into the substrate. Burrowing brings its own complications of clogging and smothering for larger animals. In slower-flowing reaches of the river, burrowing mayflies and dragonflies are often bristly, fringed with feathery hairs and spines. These projections may prevent debris smothering the animals, keeping a narrow clear volume of water around. Gills are held under or over the abdomen rather than directly out. The purpose may not be so clear cut. Burrowing species often appear festooned with debris clinging to these spines when caught in samples. Perhaps out of their natural environment the spines help with camouflage.

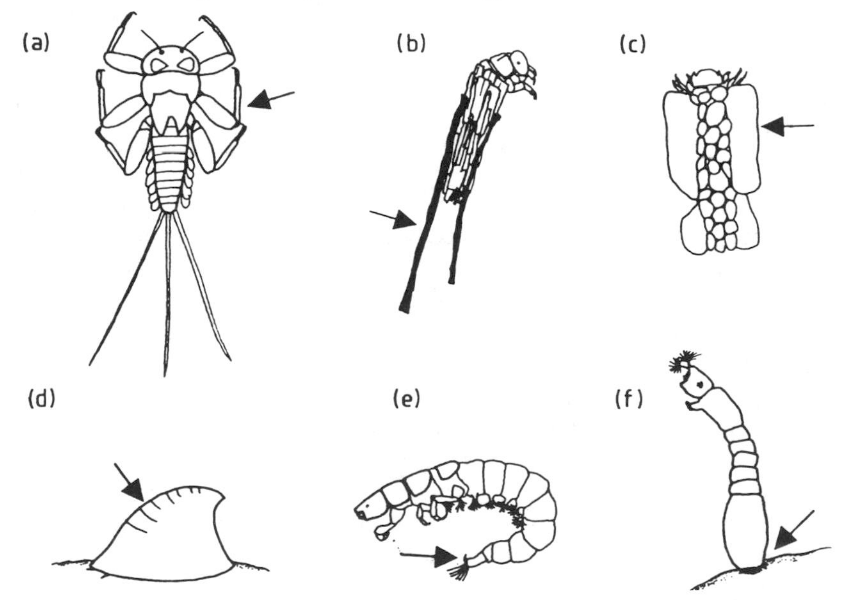

Fig. 3.6 Adaptations to the lotic environment. (a) Mayfly larva, Ecdyonuridae. The larvae are very flattened to live in the boundary layer over the substrate. The sculpted edges can be pressed down on the surface. (b) Caddis larva, Limnephilidae. The long twigs attached to the case act as vanes and keep the case aligned in the current to offer least resistance. (c) Caddis larva, *Silo* sp. Larger pebbles are lashed to the case so that it rapidly sinks if dislodged. (d) River limpet, *Ancylus fluviatilis.* Large 'sucker' foot keeps a firm grip and shell is streamlined. (e) Caddis larvae, *Hydropsyche* spp. Larvae construct retreats among stones. Cling to silk strands, especially with claws on hind prolegs (f) Blackfly larvae, Simulidae. Attached to rocks by hooks, on end of abdomen held in silken pad spun on to rock. Fan-like antennae filter food from current.

Plants must also cope with the current. Macrophytes show adaptations every bit as advanced as the animals. The form of whole plant species is often highly streamlined, at the level of individual leaves and shoots through to the whole growth form of the clump. The water crowfoots (*Ranunculus* spp.) are fine examples. Leaves and shoots are reduced to narrow filaments which are in turn sheathed with mucus. Mucus does not occur in lentic species and may resist osmotic strains (Arber, 1920). Again nature may be using the same device to solve several problems. Plants stream out in the current, the clumps forming trailing plumes. Many less clearly adapted species show phenotypic changes, heterophylly, submerged individuals often shorter and the clumps sculpted by flow while nearby plants in slacks and backwaters are upright and emergent. Whole clumps of macrophytes cannot up and move in the way animals do but many species are adapted so that dislodged and broken parts can root and grow if they are washed into safer waters. Many show annual cycles of growth followed by destruction but the remnant, broken portions are able to re-establish. Microscopic algae show responses to flow at their own scale, especially in the positions of different species on substrate (Fig. 3.7).

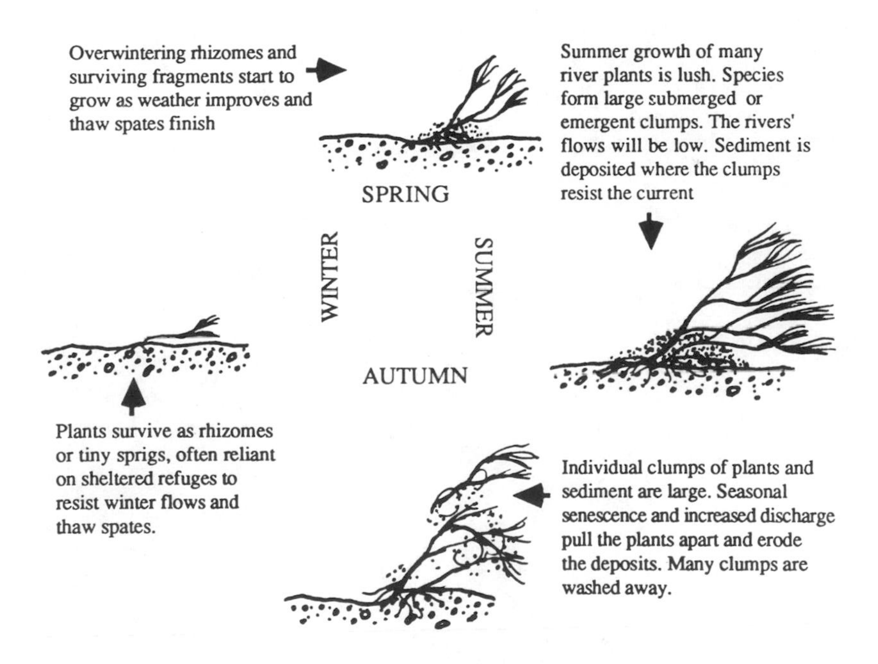

Overwintering rhizomes and surviving fragments start to grow as weather improves and thaw spates finish
Summer growth of many river plants is lush. Species form large submerged or emergent clumps. The rivers' flows will be low. Sediment is deposited where the clumps resist the current
SPRING
WINTER
SUMMER
AUTUMN
Plants survive as rhizomes or tiny sprigs, often reliant on sheltered refuges to resist winter flows and thaw spates.
Individual clumps of plants and sediment are large. Seasonal senescence and increased discharge pull the plants apart and erode the deposits. Many clumps are washed away.

Fig. 3.7 Annual macrophyte clump growth and destruction cycle. The lush summer growth during low flow coupled with sedimentation around the clump and invasion of additional species results in vulnerability to winter spates. Large clumps are washed out and regeneration begins. This causes very dynamic habitat features in many channels.

Lotic life does not have the dormancy and aestivation problem to the same degree as temporary lentic habitats. Harsh seasons are overcome as eggs, small larvae or minimal growth. Nymphs of some temporary stream stoneflies aestivate as inactive, fat-laden instars. Behavioural patterns for recolonization are complex, partly to compensate for the numbers washed downstream and out of their preferred reaches. These are discussed further in community dynamics, as life history rather than morphological adaptations.

4 Plant life

The verdant plants that spill over many waterways, the tangled trails of river weeds, the whispering reed beds or malignant slime that surface on a garden pond all conjure up images of freshwaters focused around the plant life. Water is a dominant influence on the lives of terrestrial plants so those adapted to freshwater gained a freedom to develop with an abandon denied to terrestrial relatives. The buoyant, transparent medium opens up a new habitat, the open water column, an environment the algae have made their own. The combination of water and nutrients makes some freshwater plants immensely productive, their bounty greater than most other ecosystems including even intensive farming. This productivity then supplies the wider system as plant materials are eaten and food webs ramify. Yet mysteries and dangers remain., The freshwater habitat is good for plants but there are far fewer aquatic macrophyte species than terrestrial. Too much of a good thing, especially key nutrients, can be as bad as too little, causing imbalance and degradation. Freshwaters rely largely on detritus or algae as a good source. The living macrophytes are apparently seldom eaten. The enchanted, ephemeral beauty of freshwater plants veils many secrets.

4.1 Macrophytes

4.1.1 Classification

Macrophytes is a useful term, devoid of any precise taxonomic meaning but covering all the large plants visible to the naked eye, not only the flowering plants (Angiospermae). Included are algae such as stoneworts (Charophyceae) and blanket weed (*Cladophora* spp.), mosses and liverworts (Bryophyta) and ferms (Pteridophyta). Such a farrago is ecologically useful because this assortment of plants grows together, responds to the same environmental constraints and changes, lends structure and architecture to the water and fuels food webs. Ecologically useful descriptions and classifications need not be solely taxonomic, the more so given the plasticity of growth forms.

Their growth forms, zonation and community patterns are also useful and describe not just the plants but much else about the form and condition of the environment. Hutchinson (1975) and Sculthorpe (1967) provide comprehensive, taxonomic reviews. There have been repeated attempts to devise a coherent classification scheme for aquatic macrophytes. Hynes (1970) picks out three approaches:

1. *Spatial.* Categories based on plant position in the water column and nature of attachment to substratum.
2. *Morphological.* Categories based on growth form and similarity of shape, structure and habit.
3. *Survival.* Categories based on survival adaptations for unfavourable seasons.

The spatial and morphological approaches are closely allied and can be combined. Very minutely divided schemes do exist but difficulties arise due to the plasticity of form and amphibious habits. The following is a synopsis of typical spatial and morphological schemes that make ecological sense and are useful for practical purposes.

MACROPHYTES ATTACHED TO THE SUBSTRATE

Emergent. Plants of which aerial structures are ordinarily produced. All produce aerial reproductive structures. Many species grow well on exposed substrate and survive completely submerged. This category typically includes the fringing reeds and grasses, for example reed (*Phragmites communis* Trin.), reed sweet grass (*Glyceria maxima*), club rush (*Typha latifolia*), bur-reeds (*Sparganium* spp.), and herbs, for example forget-me-not (*Myosotis scorpiodes* L.), water cress (*Nasturtium officinale* R.Br.), brooklime (*Veronica beccabunga* L.).

Floating leaved macrophytes. Plants with at least some leaves floating at the surface, attached by stems to the substrate. Many also have submerged leaves. The floating leaves usually have upper surfaces adapted to withstand exposure. Submerged leaves are often different in form from the floating ones. Reproducive organs can be emergent, submergent or floating. Examples are water lillies (*Nymhaea, Nuphar*) and *Potamogeton natans*. Some herbs may adopt this form. Amphibious bistort (*Polygonum amphibium* L.) commonly forms floating leaved mats, though it also grows completely emergent.

Submerged macrophytes. Plants that grow primarily under water, though some can resist and respond to exposure. Reproductive organs can be emergent, submerged or floating. Examples include many well-known pond weeds such as *Potamogetons,* Canadian pond weed (*Elodea canadensis* Michx.), milfoils (*Myriophyllum* spp.) and hornwort (*Ceratophyllum demersum* L.).

FREE FLOATING MACROPHYTES

Plants that grow unattached to substrate, though occasionally take hold. This group is very diverse in size, form and habit. Some float on the surface with much of the emergent structures growing clear of the water (water hyacinth, *Eichhornia crassipes*), others on it flat along the surface (duckweeds, *Lemna* spp.), just below (bladderworts, *Utricularia* spp.) or change levels seasonally (water soldier, *Stratiotes aloides*).

No such scheme is perfect, but while appreciating the dangers the general patterns are useful. The ecological characters echo the zonational seres across time and place. The gross distinctions are effective for practical conservation work (Lewis and Williams, 1984). There are additionally plants characteristic of the riparian zone and different wetlands. Lewis and Williams (1984) and Brooks and Agate (1981) provide good practical classifications for water, riparian and wetland habitats. Raven (1984) provides a classification integrated with a review of physiological and biochemical mechanisms of the plants (Fig. 4.1).

Plate 12 Submerged macrophytes: water milfoil, *Myriophyllum spicatum,* on the left and water soldier, *Stratiotes aloides,* on the right, growing abundantly in restored broadland. The *Myriophyllum* is loosely rooted, the *Stratiotes* free-floating. (Photo: Michael Jeffries)

Plate 13 Floating and emergent macrophytes. The water lily, *Nymphaea alba,* and the emergent bogbean, *Menyanthes trifoliata* (L.). (Photo: Michael Jeffries)

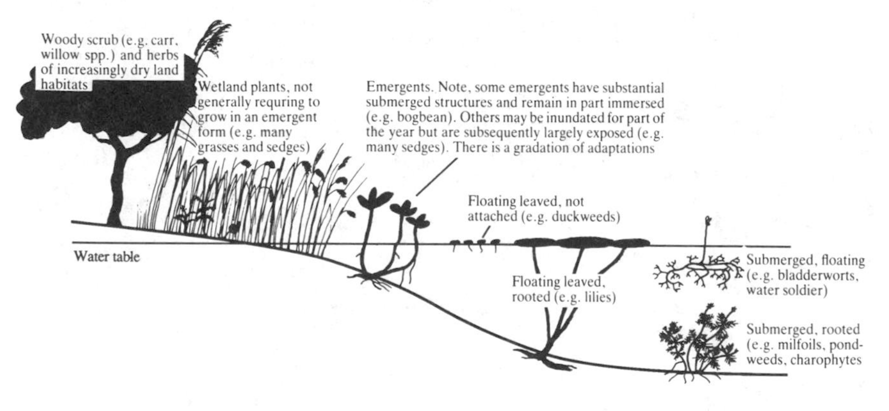

Fig. 4.1 Ecological classification of aquatic macrophytes based on position and growth form. These types would be typical of many temperate lentic–wetland habitats during summer. Note that water table and inundation will vary with season. The position of each type along the gradient from open water to dry land is also an instantaneous picture of the succession pattern of the hydroseres.

TAXONOMIC CLASSIFICATION

Taxonomically the macrophytes comprise four groups.

Macrophytic algae. Many filamentous forms function as macrophytes if they grow large but one group in particular, the Charophyceae, perform as other macrophytes. They have a stout complex architecture and are rooted by unicellular rhizoids. Species are few but found in many habitats, for example stonewort (*Chara vulgaris*).

Aquatic bryophytes. The mosses and liverworts, are ancient orders of non-vascular plants. Their life histories and structural limitations to water relations mean that many require moist habitats. Totally aquatic environments are suitable but other macrophytes are often superior competitors. Bryophytes lack true roots and many encrust substrates to compete with the periphyton. Common bryophytes are seldom dominant except in a few specific habitats; these are fast, spatey headstreams, acidic wetlands and deep, soft water lakes. In all three, angiosperms are limited by physiology and morphology. Mosses have assumed a practical importance in work on metal pollution as indicators and for bioassays.

Mosses are the only regular macrophytes of upland headstreams (Haslam, 1978). Unstable substrate, spates, acidity, low nutrients all discourage angiosperms. The unsilted substrate and turbulent water rich in CO_2 (bryophytes have primitive photosynthesis and are unable to use HCO_3) benefit the mosses. Their tough holdfasts grip the rocks directly. In cascades and waterfalls liverworts and amphibious mosses thrive. These cushions of vegetation harbour distinct communities of animals. Abrasion and smothering are dangers. The willow moss (*Fontinalis antipyretica* L.), also occurs commonly in slower, richer waters even in lentic habitats such as canals (Murphy *et al.*, 1981) and lakes (Jupp and Spence, 1977).

Unproductive, cool soft water lakes seem to have a bryophyte flora. They are able to survive at depths beyond the angiosperm light limit (Light, 1975) and barren waters also discourage competitors.

Acidic habitats, especially wetlands and lentic waters, have a typical flora dominated by mosses of the genus *Sphagnum.* They are acidophilic and further enhance acidity by fixing ions that buffer the habitats. The mosses maintain their own microhabitat and the longer term effect, as layers of moss accrete to form peat, is physically to alter the environment with blanket and raised bogs developing. Acidic and dystrophic lakes are commonly fringed by a halo of sphagnum growing out over the water as a floating rim. This may eventually grow over the water column forming a quaking mat. The build-up of peat is phenomenal but few nutrients are recycled. A distinctive associated fauna occurs, perhaps due to their acidity tolerance alone but perhaps also as a result of unsuitable detritus food inhibiting herbivores. Predatory Coleoptera, Hemiptera and Odonata dominate.

Two free-floating liverworts, genera *Ricciocarpus* and *Riccia*, occur in still, base and nutrient-rich waters. Both have global distributions.

Pteridophytes The ferns and their allies do not include many aquatic taxa but those which are typically so are the quillworts (Isoetidae), horsetails (Equisitales) and floating ferns (*Azolla* and *Salvinia*).

Angiosperms The remaining macrophytes are mostly angiosperms.

Although communities of macrophytes differ in species around the world, the patterns and underlying processes are universal. Light, nutrients, substrate, competition affect them all but the forces dominating flora in lentic and lotic waters differ. Again, rather than treat the plants as taxonomic labels, an understanding of their patterns and changes as whole communities is best.

4.1.2. Lentic macrophyte communities

Lakes and ponds commonly have a zonation of communities around them. This zonation reflects physico-chemical conditions, especially depth, at an instant but also the dynamic pattern, changing with time as the water basin fills up and increasingly emergent seres encroach and smother. The ecological classification of plant types given above nicely describes these communities.

Plate 14 A small stream (Braid Burn, Midlothian, Scotland) choked with water crowfoot, *Ranunculus aquatilis* (L.). During the winter months this is a fast-flowing stream and holds brown trout, bullheads, *Cottus gobio* (L.) and three-spined sticklebacks, *Gasterosteus aculeastus* (L.), in the quiet backwaters. (Photo. A. Grandison)

Plate 15 Emergent and wetland hydroseres. Herbs, water forget-me-not, *Myosotis scorpioides* (L.), and reed sweet grass, *Glyceria maxima*, alongside a canal. (Photo: Michael Jeffries)

Most lakes, unless shallow throughout, have an open central area. Submerged plants may grow out beneath as a sward or central substrates remain bare if too deep. In very sheltered locations free-floating vegetation may choke the centre. In shallower water the submerged sere is added to by floating leaved rooted plants. An evident ring of lilies is a common sight. Next emergents appear and within this sere a transition to wetland occurs, typified by sedges and reed swamp. In turn some tolerant trees, alder and willow, invade the mix of scrub and wetland called carr. The change is spatial along the shallow gradient and temporal as basins fill and gradients alter. This example is very neat and too perfect. Different basins have different successions. Species differ, complete seres may be absent, young basins have no obvious succession, old basins now all carr. Even within one basin neat seres are disrupted as erosional, slope and substrate character alter the precise species involved. Succession may be obvious and rapid in parts, absent elsewhere (Spence, 1964). This has been largely attributed to the local silting impact of river inflows (Sculthorpe, 1967). During any one basin's succession species increase and decline, arrive and vanish within the general seral shift (Macan, 1977). Four main factors vary the pattern:

1. Basin form, especially the original slope and depth.
2. Material forming the basin (e.g. bare rock, sand).
3. Aspect, especially the prevailing wind direction.
4. Size of the lake, which will affect the fetch of the waves.

Successions that fit the classifical pattern are hard to find, partly because of losses and degradation of freshwaters, partly because they often don't exist. The following example is drawn from the Insh Marshes, a Royal Society for the Protection of Birds reserve which is a huge tract of seasonally flooded wetland in the Spey valley in Scotland. Habitats range from open lochans to encroaching terrestrial vegetation. The open lochans commonly contain submerged species such as the milfoil, *Myriophyllum alterniflorum* DC and *Potamogeton polygonifolius* Pourret with, nearer the shore, floating leaved species such as *Potamogeton natans* and the lilies *Nymphaea alba* L. and *Nuphar pumila* (Timm)DC. Around the edges emergents such as horsetails, *Equisetum* species and the bogbean, *Menyanthes trifoliata* L. are common. The open waters are surrounded by dense sedge, *Carex*, swamp, often still very wet in summer, but grading towards drier soil, which has no water cover in summer and is covered by reed, *Phragmites communis, Phalaris arundinacea* L. and other tall grasses. Finally, dense stands of willow (*Salix* spp.) are encroaching, forming typical carr habitat. It is probably impossible to find one simple, linear gradient along which all these stages develop but they form a more haphazard mosaic as smaller scale topography features in the marsh intervene (Fig. 4.2).

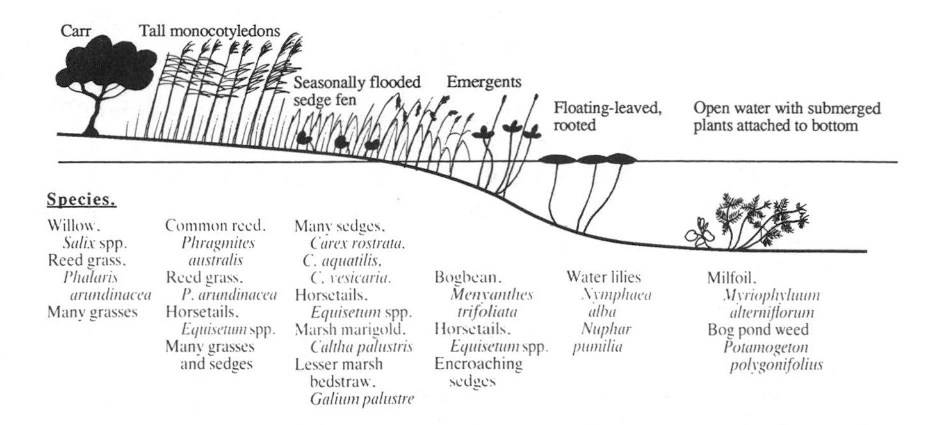

Fig. 4.2 Hydroseres surrounding oligo–mesotrophic lochans on Insh Marshes, Speyside. The neat succession shown is idealized and perfect rings of each hyrdosere do not occur around each lochan. However this is the dominant pattern, with local microterrain features adding a patchiness to give a mosaic, especially in the drier areas. The site is shown in Plates 16 and 17.

SUCCESSION PATTERNS AND THEIR CAUSES

So succession varies from basin to basin. There is an interplay of depth, hence available light and stratification, and substrate, affecting slope, composition, heterogeneity, stability and nutrient content. The wind–wave regime controls erosion and sedimentation. There are also strong biotic interactions, largely competitive, one or other species losing out, but sometimes facultative, one or both are helped by the presence of the other. The balance of dominance of these different influences varies from basin to basin but a general order can be worked out. Note that the factors operate within the broader biogeographical context; the absence of a plant from a basin's zonation in a continent from which it is entirely absent is nothing to do with the depth, substrate or competition.

Many measures of the depth of individual species or communities have been made and two general trends are consistent. Firstly, angiosperms rarely descend to more than 9–11 m, while algae and bryophytes are known to exist much deeper, mosses to 20–30 m, Charophyceae 10–40 m, *spirogyra* (a filamentous algae) 52 m. Secondly, despite known depth tolerances of angiosperms, many species seldom occur in zones neatly explained by light alone. Other factors appear to limit distributions, within the overall light tolerance bands. Perhaps a hierarchy of processes can be teased out. Recent acidification resulting in clearer waters in some lakes has increased known depth records (Singer *et al.*, 1983).

Depth in water causes very rapid changes. Hutchinson (1975) points out that a 15 m descent produces greater changes in light and pressure than a 150 m change in a terrestrial system. Sharp thermoclines can cause big changes across less than 1 m. There are links between the lower limits to macrophyte growth and turbidity and rankings of individual species tolerances can be worked out. Interspecific differences in photosynthetic physiology are detectable. Pigment differences occur. The deepest macrophytes probably represent species limited by light. In shallower water their absence and restrictions of other species probably reflect other factors which become predominant once light levels are generally good for any species. Pressure also increases with depth. For every 1 m down an additional 0.1 atmosphere is added, on top of the air's 1.0 atmosphere. So at 1 m depth pressure = 1.1 atm; at 5 m = 1.5 atm; at 10 m 2.0 atm. Experiments have shown growth and morphology changes at pressures of 1.5–1.6 atm and severe inhibition at 2.0 atm in several species. The 2.0 atm point is roughly coincident with the 9–11 m angiosperm limit. At this pressure lacunae become waterlogged, respiration, photosynthesis and nutrient metabolism are all adversely affected. Algae and bryophytes lack such lacunae. Their deeper penetration may be a result of pressure factors. The changes with slightly increased pressure may still be enough to shift the competitive edge from one species to another. So plants may not reach their absolute limits of pressure and light tolerance as they are also at a competitive disadvantage at their lower tolerances to other, better adapted species.

Basin substrate factors include slope profile, exposure to wind and wave action and substrate composition. Slopes of different steepness have different zonation, largely the result of differing sedimentation. The gradients also mean that depth changes occur over different lateral distances from the shore so zones can be truncated or stretched. Exposure to wind and waves is often an obvious problem. Exposed shores of large lakes often lack any macrophytes. Shallow lakes, where wave fetch is reduced, and sheltered bays or the lee of an island maintain plants. Emergent vegetation is very sensitive to exposure and will be better adapted to sheltered shores or restricted to havens. There are interspecific differences to erosion and sedimentation. Tough monocotyledons, for example bulrush (*Schoenoplectus* spp.), *Phragmites* and *Typha* are more resistant than herbs. The three monocotyledon genera also differ between themselves, and are listed in order of increasing vulnerability. These are important points for management of bank and shore erosion. Emergents facilitate the occurrence of small free floating plants sheltering among the stems.

Floating leaved plants suffer battering and damage. Generally, smaller leaves, for example *Potamogeton natans*, are more resistant than larger, for example lilies, *Nymphaea* and *Nuphar*. For the submerged plants it is the erosion/sedimentation on the basin slope that becomes important. At the water's edge of exposed lakes an erosion zone occurs. Frequently the smaller material is washed out leaving just large boulders and bedrock. Below this is an area of mixed erosion and deposition, further down sedimentation. Isoetids are particularly adapted to these exposed shores, though each species shows slight differences. Shoreweed, *Littorella uniflora* (L.), typically colonizes highly disturbed water, quillworts, *Isoetes* spp., unable to adjust rooting depth, prefer neither eroding nor sedimenting areas. Commonly associated plants are *Nitella* (a charophyte) and *Juncus bulbosus* (L.) (a rush), both preferring the deeper sedimenting zone. Other macrophytes prefer less exposed shores and more substrate. The substrate's influence can be ordered as erosion/sedimentation> composition (texture, heterogeneity)> organic matter, nutrients and oxygen (Fig. 4.3).

Competition is important. The Charophyceae and mosses grow deeper than the angiosperms, and the isoetids and associates of exposed shores are not limited to poor environments. They can thrive in richer, sheltered sites. Their absence, or restriction to small areas in less stressed waters is due to competition. It is a typical pattern in ecological communities to have a gradient from harsh conditions, tolerated by rather few highly adapted species, to clement conditions dominated by other species, unable to survive the harsher zones but out-competing the tolerant species under the easier conditions. Freshwater seres can be largely monoculture. Aggressive vegetation propagation allows rapid dominance by the first species to establish, a priority effect. Subsequently the physical structure of emergent stands, dense swards or tangles, roots and rhizomes make it difficult for other plants to invade. Emergents shade out plants below, floating leaved or free-floating species shade submerged ones. Competition can work in two

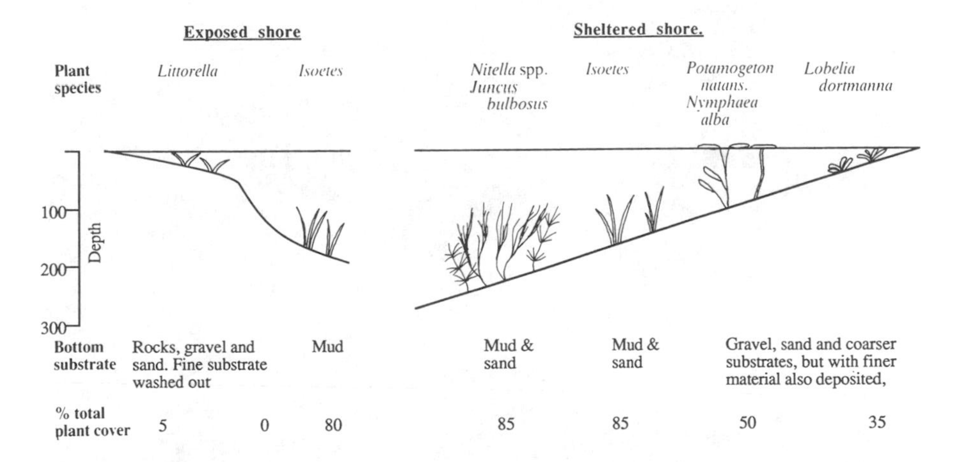

Fig. 4.3 Typical plan communities on exposed and sheltered shores of oligotrophic lochs. Only *Littorella* can survive on the exposed shore, where wave action disturbs the sediment, damaging plants, preventing rooting and removing richer, finer substrate. (After Spence, 1964)

Plates 16 and 17 Hydrosere succession at Insh Marshes, Inverness-shire, Scotland. Plate 16 Open water with submerged and floating plants and encroaching sedges; Plate 17 Sedge fen being invaded by reeds and willow carr (see Chapter 4). (Photo: Michael Jeffries)

ways: interference, in which there is direct interference between competitors; and exploitation, accruing the limited resources faster than the competitor, but without any direct action. Plants employ both. Rapid growth will cause direct interference by smothering and shade. Chemical secretions also fall into this category. Use of secretions specifically to attack competitors is termed allelopathy. Equally plants have differing abilities to utilize nutrients, so exploitation competition also happens.

As a general rule limitations of light and pressure form the extreme boundaries to species and community patterns. Erosion/sedimentation/exposure dominate where light and pressure are adequate. On substrate of equal exposure the precise composition, heterogeneity and nutrient content become important. Across all this hierarchy competitive interactions may resolve the final dominance.

4.1.3 Lotic macrophyte communities

Lotic macrophytes show similar general traits to those of still waters with typical communities of river types, changes with time and community dynamics responding to water substrate, nutrients and biotic interactions.

The dominant general pattern is of longitudinal zonation from headstream to lowlands reflecting the changes described in Chapter 3. Plants respond to channel width and slope, flow patterns, substrate and nutrients, just as lacustrine species do. In Britain typing of rivers using macrophytes has been extensively developed for conservation and management. Such schemes have identified at least forty river types and even a simplified version relies on ten main types (Holmes, 1983). Haslam (1978) provides very detailed descriptions of plant macrophyte communities, also extended recently to include Europe.

A description of three basic types is a useful foundation:

1. *Mountainous, resistant geology.* Fast spatey flows, coarse substrate and nutrient poor. Headstreams may be completely bare or limited to bryophytes. Surrounding bog vegetation may encroach, for instance sedges (*Carex* spp). Fringing herbs become common downstream with occasional channel tufts of milfoil, *Myriophyllum alterniflorum* and crowfoot, *Ranunculus fluitans* Lam. All these are washed during spate. Occasional slack stretches harbour fringing monocotyledons and bur-reeds, *Sparganium*, in the channel. Overall, limited diversity, few channel species and regular washing out.
2. *Upland, oligo–mesostrophic rivers.* Many upland systems are not so steep and severe. Spates are less characteristic, sediment accumulates and trophic status rises, partly due to human interference and inputs. These often have well-developed species-rich fringing herb communities. In the channel *Ranunculus* and starworts, *Callitriche*

spp., penetrate far upstream. Fringing monocotyledons such as *Phalaris* and *Sparganium* species are common. Such upland rivers often grade into the third general type but there are many examples of lowland, alluvial systems that are lowland throughout.

3. *Lowland, often alluvial, eutrophic.* The most spectacularly verdant forms are those rivers on chalk. Steady flow from aquifer sources, rich in nutrients and shallow with a stable substrate, they have many herbs and channel species forming dense swards throughout their lengths. Marked zonal changes may be indistinct. Other lowland rivers lack their sheer diversity. Those on alluvial deposits, especially clay, have monocotyledon choked headstreams, often dikes and ditches. Large channels with unstable substrate and often nutrient enrichment also have lower diversity. Eutrophic species are common, lilies and grass-leaved *Potamogetons*. Fringing reeds, sedges and iris project a limited herb flora that lines the banks (Table 4.1).

LOTIC PATTERNS AND THEIR CAUSES

Plant species and whole communities do show correlations with general channel width, depth and slope (Haslam, 1978). The zonal gradients reflect changes in these influences. The width–depth–slope are equivalent to the

Plate 18 Extensive *Phragmites* beds bordering the eastern shore of Neusiedlersee lying between Austria and Hungary. The ecology of this lake is described concisely by Burgis and Morris (pp. 124–7, 1987). (Photo: Derek Mills)

Table 4.1 General vegetation patterns along oligotrophic, mesotrophic and eutrophic river systems.

RIVER SECTION	TROPHIC STATUS Oligotrophic	 Mesotrophic	 Eutrophic
Headstreams	Upland, resistant geology, spatey, coarse substrate Few vascular macrophytes in channel. Mosses and algae dominate. Fringing plants include herbs and grasses, e.g. *Agrostis stolonifera and* reed grass. Common in uplands acid bog species, e.g. bog pondweed, sedges and cottongrass	Such rivers typically start from spatey upland streams and have a similar fauna to oligotrophic systems. Acidic flora generally absent and patches of herbs and grasses may thrive in protected refuges	Generally lowland rivers on soft geology or alluvium. The headstreams are often shallow gradient and nowadays often ditched and diked in areas of intensive agriculture. Many herbs and grasses flourish and can clog the streams, e.g. water cress water celery, reed grass, reeds, bur-reeds, starwort, water mint
Middle reaches	Channel macrophytes remain sparse. Fringing herbs may be abundant, washed out and recovering each year, e.g. brooklime, water celery, water cress. Mosses and algae still frequent	Softer rocks, lower gradient, deposition and richer nutrient supply all allow good vascular flora to develop. Many herbs, grasses and channel species, e.g. crowfoots, broad leaved and curly leaved pondweeds, floating and erect bur-reed, brooklime, forget-me-not, water cress	As channel size increases the herbs are replaced by genuine submerged channel species with fringing grasses, reeds and emergents typical, e.g. milfoils, floating bur-reeds, lilies, arrowhead, water plantain, reeds, reed grass, starworts. Many such rivers are now polluted with excessive plant nutrients and are turbid. In such cases channel species may be limited and algal growth abundant
Lower reaches	Lower gradient and deposition allow vascular macrophytes to survive. In channel crowfoots, and fringing herbs and grasses abundant. Mosses and algae may now be outcompeted		

basin-substrate factors in lakes and determine the flow and substrate.

The depth factors of light and pressure are only rarely of importance in rivers. Light is not a restriction except in turbid or polluted channels, typically in lowlands. Riparian shading will influence species and productivity.

Flow sorts sediment but also acts on plants directly. Leaves and shoots are battered, the load abrades the tissues. The actual pull of the water (termed velocity pull) acts in two ways. Plants present a degree of hydraulic resistance. This varies from species to species and with size and shape of plant. Bushy, erect species have a higher resistance than streamlined submerged ones. The pull also tugs at the plant's attachment to the substrate, which depends on anchor strength. This may resist flow but plants can be pulled up whole or the exposed tissues snapped off. Different plant species have different anchor strengths. Shape and toughness of tissues create differential resistances to abrasion and battering. The precise balance of species and growth forms depends heavily on the flow (Haslam, 1978; see also Fig. 4.4).

The substrate structure affects plants by the rooting or attachment it permits, vulnerability to erosion and sedimentation. As a general rule many species of fast, coarse substrate habitats root shallow as there is little in which to root. Those of deeper, slower channels root deep into the fine deposits. Erosion can dig out plants and, again, the form and depth of rooting of different species varies their vulnerability. Different substrates will also resist erosion to varying degrees, Sedimentation can be an equal threat. Tall plants commonly grow through but many cannot and are smothered.

Aquatic plants take nutrients largely from the substrate. Different catchment geologies will provide differing amounts of silt, which varies in nutrient content. Flow then determines how much is deposited. The plant

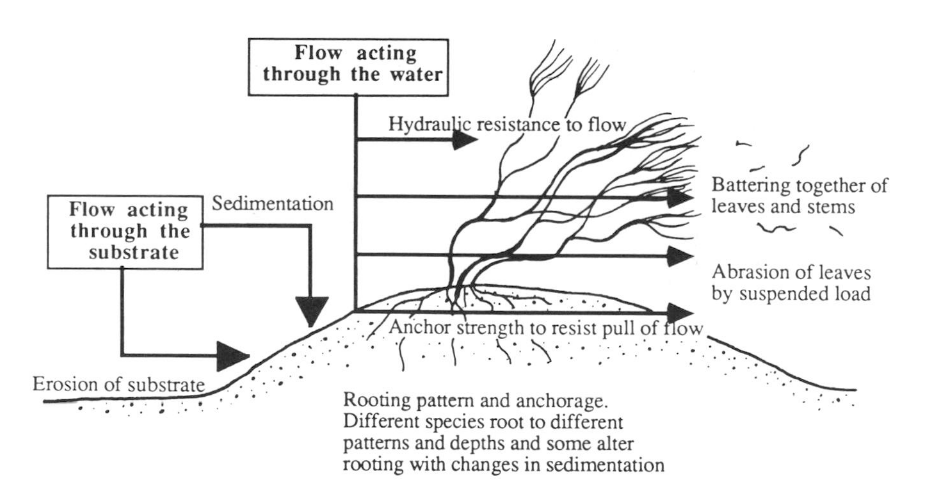

Fig 4.4 How flow affects macrophytes. Action through the water includes the physical forces that are exerted on the plant, which may uproot it whole or snap the roots and stems, plus damage from abrasion and battering. The flow also determines erosion and deposition. Different species respond differently to increasing sedimentation. (After Haslam, 1978).

communities do reflect the river's trophic status. Plants can be assigned to oligo- through to eutrophic categories and preferred ranges of ion concentrations have been worked out for some species (Haslam, 1978; Holmes and Newbold, 1984). Increasing nutrient inputs from human activity greatly shift trophic balances and certain species are regarded as good indicators of enrichment, for example *Potamogeton pectinatus* L.

Competition is as poweful a factor as it is in lakes, with the balance shifting back and forth between species in response to changes in the rivers' regimes. Several studies show changing dominance over time (Ham *et al.*, 1982; Wright *et al.*, 1982; see also Table 4.2).

The interspecific differences in flow and substrate needs, competition and seasonal cycles of growth and destruction result in a common cycle of channel plant communities. Following spates, which wash out old clumps, new plants grow from remaining tatters or colonize from havens. The new

Table 4.2 The role of macrophyte species architecture in animal community patterns. In this study animals and detritus were sampled on four plant species, figured below, and samples quantified relative to surface area of plant sampled. Higher numbers of animals are associated with higher quantities of detritus. Species with more complex architecture may trap more detritus. However, other factors such as interspecific differences in plant epiphyton, and more complex relationships between animal size distributions and habitat complexity probably also play a role. This is an elusive and fascinating topic still largely unexplored. (Data from Rooke, 1984)

	Dissected leaved plants		Broad leaved plants	
	Chara vulgaris	*Ranunculus longirostris*	*Potamogeton richardsonii*	*Potamogeton amplifolius*
No. animals m^{-2}	3815	1089	3147	1135
Dry weight of detritus ($g\ m^{-2}$)	7.25	2.09	7.44	5.28

plants grow, the clumps causing sedimentation and perhaps facilitating the growth of other species in their sheltered lee. The clumps alter the flow and erosion, which become increasingly severe between the plants gouging the channels. Eventually clumps grow too big, too exposed and are washed away in their turn. Such cycles in the channel profoundly affect the other life. Alterations, either to flow, sediment or gross channel engineering will reset the rhythms and cycles, interrupting the natural processes.

4.1.4 How macrophytes affect the environment and wildlife

Macrophytes are not passive objects, dependent on the whim of their environment. They have a profound impact (Sculthorpe, 1967; Marshall and Westlake, 1978; Carpenter and Lodge, 1986). Because of this their presence or loss, management and mismanagement present problems and opportunities.

Their metabolic activity alters the physical and chemical surroundings as in streams (Gregg and Rose, 1982). Shade, temperature, oxygen, CO_2 and pH are obvious factors that affect growth, photosynthesis and respiration and which will change at many scales of time and place (Edwards and Owen, 1962; Dale and Gillespie, 1977; Halstead and Tash, 1982). The speed of change and extreme reached are important. Besides these obvious abiotic factors probably all other physical and chemical processes are altered by plants. This can be put to use. Plant management can be used beneficially, to provide oxygen, as a sink for nutrients and even for dealing with pollutants that have little place in natural systems, for example the use of water hyacinth to strip heavy metals from water.

Plant chemistry has more direct impact. Secretions by macrophytes partly determine their epiphytic algal cover. Some species have an extensive film, others are virtually bare. Macrophytes also inhibit algae competitively, either by locking up nutrients or by direct chemical interference. More curious is the apparent inhibition of some animals. There is evidence that macrophytes inhibit invertebrate development (Gonzalves and Vaidya, 1963). Macrophyte–insect interactions are an intense area of research in terrestrial systems but largely unexplored in freshwater. Living macrophytes are often regarded as an under-utilized food source. Although freshwater plants do show evidence of defensive chemistry, measured by chemical assays (Ostrofsky and Zettler, 1986) or by palatability tests with foliage eaters (Jeffries, 1990), chemical defences do not seem a likely explanation. Certainly there are plenty of terrestrial insects that can cope with hideous arsenals of plant weapons. The answer may lie at an evolutionary level, the major macrophyte taxa have not invaded freshwaters. This area remains a mysterious and enticing area of research.

The physical structure of plants is important. The zonation in lakes and clumps in rivers provides vital habitat and studies of animal communities show distinct associations (Dvorak and Best, 1982; Cyr and Downing, 1988). The plant structure, the architecture, regardless of its actual possible food role, counts. Habitat structure is another burgeoning area of research, especially with developing links between advanced mathematical measures of structure (e.g. fractals) and ecology. Precise animal–plant architecture studies will be very exciting given the array of plant shapes from the simplicity of a lily stalk and leaf to the feathery weft of a milfoil clump. The structures, both submerged and emergent, are important as shelter for adults and juveniles. Proper maintenance of vegetation is vital to fisheries and for amphibians. The introduction, management and removal of macrophytes is a useful practical tool.

A final aspect often overlooked in the dry output of science is that these plants are immensely beautiful. The extravagant, ephemeral plants of freshwaters are a hugely appealing sight. The importance of this for conservation and amenity, as a potent argument for their proper protection, should not be treated as a trivial nicety.

4.2 Microflora

The divide of macrophytes and microflora is one of size and life style, cutting across taxonomic classification. The groups have an ecological validity, sharing ways of life, resource needs and community patterns. The microflora consists of tiny, often single-celled organisms, many of them far from being genuinely 'flora'. It includes bacteria and the allied blue–green algae, true algae (a disparate group itself) and fungi. For all their small size as individuals their productivity is vital to sustaining life and microflora do much to mould the overall form and variety of freshwaters.

The taxonomy and classification of these organisms is complex. Divisions within the general grouping of the algae represent divides greater than those between larger taxa that we take for granted. Good reviews of algal taxonomy and ecology can be found in Round (1981) and Bold and Wynne (1985). Another useful classification is based on their ecology, where they live, what they do. The microflora can firstly be divided into two major types: the phytoplankton, normally living in the open water column, and the periphyton, living attached to a substrate.

4.2.1 The phytoplankton

The pelagic zones of lakes and ponds are worlds dominated by the phytoplankton. Their huge population, seasonal cycles and nutrient processing are spectacular. The plankton consists of ten main taxonomic groups. These are: viruses, bacteria, blue–green algae (Cyanophyta), green algae (Chlorophyta), golden-browns (Chrysophyceae), diatoms

Plate 19 Riparian vegetation along the banks of Afon Praw in Wales. Among the bankside plants present are the iris, *Iris pseudacorus* L. and the marsh cinquefoil, *Potentilla palustris* L. The branched bur-reed, *Sparganium erectum* L. and horse tail, *Equisetum,* line the stream margins while the broad-leaved pondweed, *Potamogeton natans* L., covers much of the stream surface. (Photo: K. Morris)

(Bacillarophyceae), cryptomonads, dinoflagellates, Xanthophyceae and fungi. They are a cornucopia of shapes and sizes. Size is used as a classification which has much practical merit. The smallest, less than 10 μm, make up the ultraplankton. Those 10–50 μm are called nannoplankton and those 50–500 μm the microplankton. Those larger than 500 μm are macroplankton. Their ecology also depends on feeding habits. Fungi are heterotrophic, they require food from other organic sources, dead or alive, which they digest and ingest. The algal groups vary from those that are almost entirely autotrophic (manufacture all requirements from basic elements and photosynthesis) through to others that are almost entirely heterotrophic, engulfing food particles, with all grades in between and many able to switch from one mode of feeding to another.

PHYTOPLANKTON POPULATIONS: PATTERNS AND DYNAMICS

The pelagic world is not the uniform environment we see. Currents and water masses, nutrients and chemistry weave an invisible heterogeneity. Phytoplankton populations in a lake may be very patchy. Reynolds (1987) gives an integrated synopsis of environmental events and phytoplankton responses showing heterogeneous patterns across scales of minutes to decades (Fig. 4.5).

The total lake populations show seasonal dynamics with a succession of species. The changes in phytoplankton populations are important in regulating the whole pelagic ecosystem. The general pattern for temperate lakes runs as follows. Following the cold, poorly lit winter months day length increases, water warms and winter mixing will have recirculated nutrients from the enriched deep water. Conditions are ideal for algal growth and a spring bloom breaks out, often dominated by diatoms. The spring bloom may crash as quickly as it grew. Nutrients are used up and thermocline formation prevents additions from deep sediments. The loss of diatoms is accelerated due to a specific need for silica for their cell walls. Green algae also respond to the spring flush but tend to dominate later in the summer once diatoms have declined. A sequence of species dominates in turn. Diversity may be high, though numbers of any one species are low. Small Cryptophyceae and Chrysophyceae thrive. Blue–greens also bloom in summer. Tolerant of stagnant, anoxic conditions they have the added advantage as many are able to fix atmospheric nitrogen and release toxins to suppress competitiors. Autumn, with thermocline turnover and changing light conditions, often prompts a small bloom of diatoms before winter ushers in a lean time of inactivity. Such patterns are common but do not be lulled by their simple attractiveness. Bailey-Watts (1982) presents data for a study over many years in Loch Leven, Scotland, showing that while the overall species present do not change, their relative abundance and peaks reached vary greatly from year to year (Fig. 4.6). World-wide successional patterns are described in Munawar and Talling (1986).

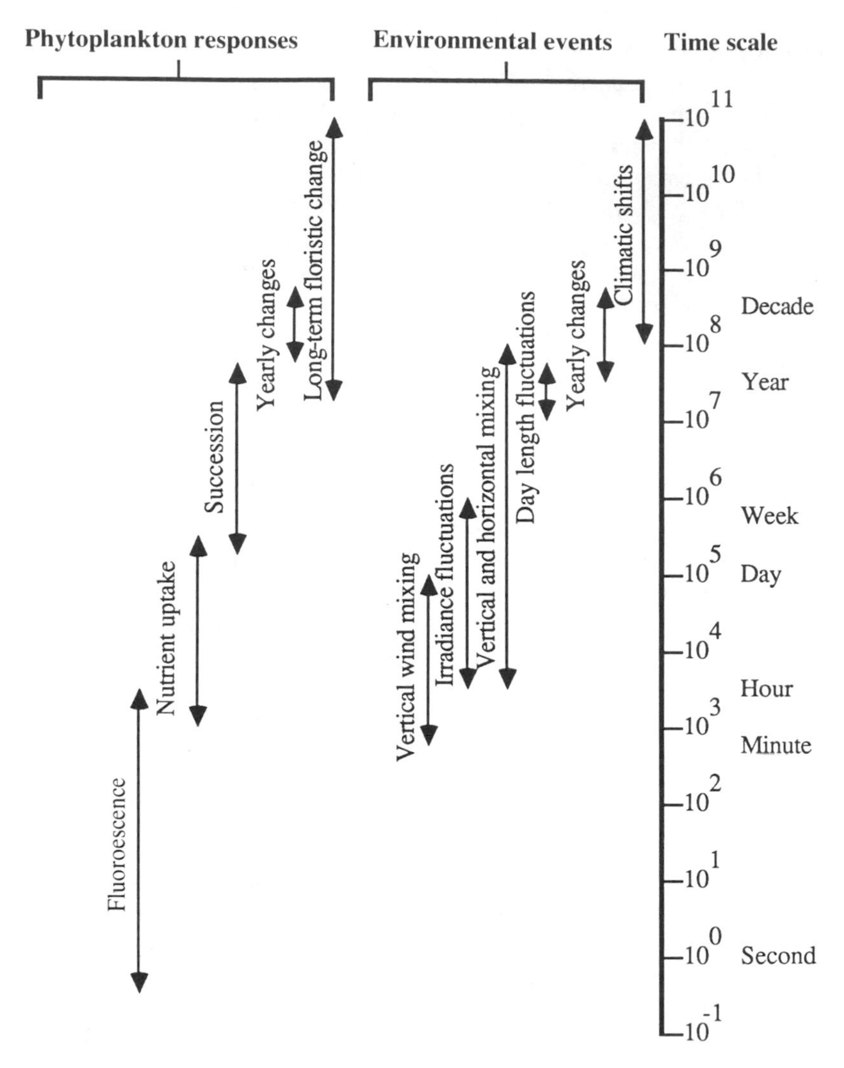

Fig. 4.5 Time scales for selected environmental events plus the related responses and response periods for phytoplankton. (After Reynolds, 1987)

Much research has been devoted to unravelling the controlling forces behind these community dynamics. They fall into three broad categories (Hutchinson, 1975):

1. *The physical environment:* temperature, light, turbulence, stratification. These are hydrodynamic factors.
2. *Nutrients:* hydrodynamic and nutrient factors are 'bottom-up' controls, structuring the communities from the base of food webs and as physiological limitations.

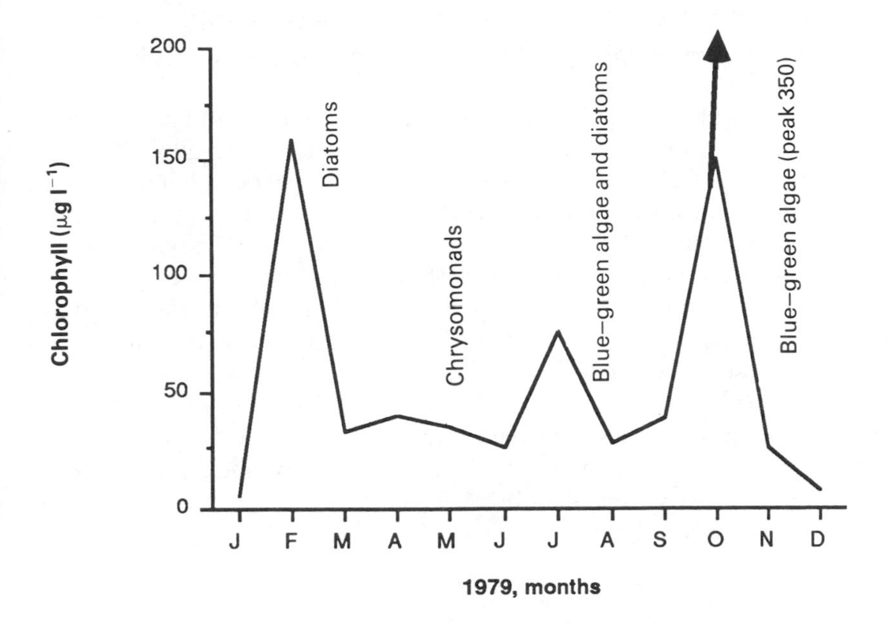

Fig 4.6 A typical annual phytoplankton succession in a temperate lake. A spring bloom of diatoms develops as climate improves but crashes as nutrients are exhausted. Summer is dominated by small chrysomonads and later blue–green algae many of which can use atmospheric nitrogen so flourish when dissolved nitrogen is very low. The abundance of algae was measured using the concentration of chlorophyll a, given as g l^{-1}. (After Bailey Watts, 1982)

3. *Biotic factors:* competition, toxic inhibition, competition and grazing. Note that while herbivory and predation tend to mean very different things in common use, they both essentially mean one thing eating another. Ecologically, predation is often used to cover both. Given the non-plant nature of so much of the phytoplankton predation is a good term.

   Biotic factors are commonly 'top-down' such as predation, parasitism and disease, exerting their influence down through a food web.

Algal populations change with changing physical conditions but the actual effect is perhaps due to physical conditions altering nutrient supply. Physical changes may act more directly as cues. Increasing light and temperature herald spring. Many algae have preferred thermal environments. There are warm water and cool water species, not just in the obvious global sense but living in the same lakes, thriving best at different seasons. Thermocline formation and loss precipitate change. Responses to nutrient levels have been analysed in laboratory cultures and field studies with enrichment. There are many examples of algal populations changing coincidentaly with nutrient levels. Although any essential nutrient, from simple ions to vitamin complexes, in short supply can inhibit populations, nitrogen and phosphorus are the two keys.

Biotic controls are tricky to sort out. Competition seems to occur, mediated through the nutrients and abilities of species to glean them. A more direct competition, by production of toxins has been demonstrated. Control by grazing and parasites has been less well explored. Many filter feeding zooplankton have phenomenal filtering powers. They can be size selective. Different algae possess different defences and even survive ingestion. So not only will total populations be altered by grazing but so will species composition. Some examples of heavy fungal attack, coincident with bloom collapse, have been described.

The tangled ecology of phytoplankton remains a challenging topic. Reynolds (1987) suggests that hydrodynamic factors are the driving force behind phytoplankton communities, with the patterns and structure of the communities weakly organized but evidence for the importance of biotic pressures such as grazing is also strong.

4.2.2 The periphyton

The taxonomic composition and classification of the periphyton is broadly similar to that of the phytoplankton, with considerable overlap even at the genus level. An additional, useful division of periphyton is based on their substrate: epiphytic on plants; epilithic on rock; epipsammic on sand; epipelic on fine sediment; and epizooic on animals. Periphyton flourishes in most lakes, within the limits of the photic zone but also in rivers. The macroflora are tiny but they are adapted to cope with just the same constraints as macrophytes. In rivers, epilithic communities exploit the boundary layer and microdistribution on rocks varies with surface topography. The communities respond to changes in flow. Epiphytic communities are often the densest periphyton. The rough macrophyte surfaces, multiplicity of shoots and leaves and lowered flow inside clumps are good habitat. Epipelic and epipsammic floras include many mobile species, both single cells and filaments creeping in response to light, often to diurnal rhythms. They can move in the shifting substrate. The periphyton of lentic water does not have flow to contend with. Instead, smothering is a greater problem. Epilithic and epiphytic flora can be profuse, though both are especially susceptible to dangers from water level changes, shading and grazing. Some macrophytes suppress epiphyton directly with secretions. Distinct macrophyte–periphyton associations form. Some periphyton are specific to macrophyte and even precise microhabitat on the larger plant. Loss of macrophytes will cause loss of epiphyton. Though epiphyton are an important food for grazers rather than the macrophytes, loss of the latter from pollution or mismanagement will still interfere with secondary production. Macrophytes that support only a poor epiphyton can be equally barren for herbivores. Sediment periphyton are affected by the substrate mobility, composition, texture, oxygenation and organic content.

PERIPHYTON POPULATIONS: PATTERNS AND DYNAMICS

Periphyton change with seasons but the neat patterns found in the phytoplankton are harder to pick out. This is partly a result of difficulties with sampling and the often more diverse community but the more complex environment results in more complex patterns. A spring bloom is common, again dominated by diatoms, but overall summer populations may steadily increase rather than crash, until autumn heralds leaner times. In the case of epiphyton the seasonal development of their macrophyte homes would be enough to cause this increase. In lotic habitats the flow of water will constantly renew nutrients and in lakes closeness to substrate may result in fewer nutrient restrictions than for plankton. Riverine periphyton is sensitive to shading.

The periphyton also respond to anthropogenic changes. The River Tweed, between Scotland and England, has recently suffered massive growths of the diatom *Navicula avenacea* apparently caused by fertilizer input from the Tweed's large catchment of rich agricultural land (Currie, 1989).

The ecology of freshwater macrophytes, phytoplankton and periphyton is complex. However the responses and adaptations of the plants are individual solutions to coping with and exploiting a shared set of problems such as flow, light and substrate. In their turn the plants pose problems and opportunities for the fauna.

5 Animal communities – the abiotic world

Unravelling the patterns of animal communities and the underlying causes is science at its most complex and exciting. Ecologists have to integrate the effects of environment with the interactions of animals and plants, and they must do all this across scales ranging from millions of years to milliseconds. The importance of this is not simply its esoteric fascination but to help us manage the landscape and waterways properly. Much ecological research has been taken up trying to demonstrate the predominance of one influence over the others, be it biotic or abiotic. The general lesson is that no single factor – climate, geology, geography, habitat, competition, colonization, predation or sheer chance – is universally the most important. All factors act to some extent (if only by being absent) in all ecosystems. Their relative importance in a particular system varies, again with changes across scales of time and space. Picking out the relative importance of different influences, at the scales you are interested in is the trick. This chapter and Chapter 6 outline the role of abiotic and biotic factors, respectively, in shaping animal communities. Examples are drawn from many and varied habitats illustrating the shifting importance in the balance of factors from habitat to habitat. This chapter is concerned primarily with abiotic influences: the water chemistry, the flow of rivers, the size of substrate. Chapter 6 looks at populations and their interactions, such as competition and predation.

For the applied freshwater ecologist the operation of these obvious factors, recorded over a season, a year or perhaps decades provides vital information. In the background though are two factors, working on a huge scale and perhaps not immediately obvious in our short-term view, but which provide the framework within which the other factors operate.

5.1 Palaeoecology and zoogeography

It is facetiously obvious that studying freshwater systems today you do not find the same species that dwelt in primordial coal forests, or, studying streams in the Northern Hemisphere you do not find the same species as in a southern rain forest, although in both cases similarities at higher taxonomic levels can be striking. Such results are so evident that we tend to ignore these large-scale patterns: the first of time, the second of geography. The distribution of animals and plants across these scales provides the pool from which local fauna and flora are drawn and in recent years considerable advances in palaeoecology have filled in much of this background (Berglund, 1986). Waterfleas (Cladocera), ostracods (Ostracoda), beetles, Chironomidae, molluscs and caddis flies have received most attention. Whole fossil communities have been described and the histories of individual waters described. Large-scale changes in the fauna can be reconstructed as communities waxed and waned over huge time periods. This provides a background against which modern community patterns must be seen. Beetles have proven a particularly fruitful group. Changes in beetle fossils have allowed interpretation of postglacial climatic changes in Britain, with retreats and advances as climate worsened or ameliorated. So the general climate across a region will determine the gross pool of species from which communities are made up. The fossil remains can often be identified with living species. In Britain, since the ice retreat 13 000–15 000 years ago the actual species present have been stable on the evolutionary scale. Angus (1983) presents evidence that species of the aquatic *Helophorus* beetles have not changed in form for 10 000 years, but geographical ranges have shifted with climatic change and communities can alter dramatically (e.g. Coope 1977, 1986). The communities we find today are just an instantaneous sample of much longer term trends.

5.2 Invertebrate communities and the abiotic environment

The patterns of time and space create the background in which the details of smaller scale abiotic and biotic processes work. It is these influences and how our management alters them that concern most freshwater ecologists. The literature on aquatic animals is vast with extensive reviews of all aspects from individual species to computer models. The aim of this book is to look at whole communities and how they function and respond. It is at this level the discussion is concentrated. Different communities (planktonic versus benthic, lotic versus lentic) are often treated separately. Evidently differences occur but here discussion of these communities is integrated, to help develop the important themes that affect them all. Where differences do occur these are pointed out. While the details may differ, for example a lacustrine plankton community may be differently affected by acidity changes than a stream benthos, the general process (changing acidity) is the same.

Rather than dividing the discussion by habitat a division has been made on different criteria. Responses to the abiotic environment are described at three levels: the individual species (autecological); communities and the aquatic habitat; and communities and the whole catchment.

5.2.1 *Individual species and their environment*

Since communities consist of individual species a look at the responses of animals at this level is a good foundation. The patterns seen at an autecological level rapidly multiply as species interact in intact communities. The ecology of many species has been extensively researched and here two taxa, *Daphnia* water fleas and *Gammarus pulex* (L.) the freshwater 'shrimp', will be described in detail. Both are widespread but patchy in their distribution, suggesting distribution does vary with prevailing conditions. Both have been well researched and there are closely related species which provide useful comparisons. A vertebrate, the brown trout, *Salmo trutta* is described later. Their autecology is particular to each species but will provide general principles of how any species responds to its environment.

The *Daphnia* are a genus of water fleas (Cladocera), crustacean zooplankters. They are crustaceans, as is *Gammarus*, but their ecology, as filter-feeding plankton of the open water is very different. Work has been carried out intensively on several species, notably *Daphnia pulex* (De Geer). However, due to difficulties of identification, the worldwide distribution of many species and morphological plasticity (i.e. the body shape of individuals of a species) can vary markedly and some hybridization description of just one species would be dubious and less interesting. The following outline refers largely to *D. pulex*, with some points illustrated for close relatives.

The *Daphnia* are a very ancient group (Benzie, 1987). 'Resting eggs' (produced to resist harsh conditions) and egg pouches have been found in Tertiary period deposits, the geological period from some 65 million years ago to the present day, that are indistinguishable from those of modern species. Australian fossils suggest origins in at least the Permian period, over 230 million years ago, when most land was joined together in one vast continent, Gondwanaland. When this split up the genus would spread to all the different parts, hence the worldwide distribution. Many species are cosmopolitan, that is found widely across the world, from east to west and in both hemispheres. The result is that species complexes occur around the world. These are extremely similar types that might be genuine species but could be zoogeographical varieties of essentially one species. Some species occur in both temperate and tropical lakes suggesting basic physico-chemical tolerances are wide.

The pelagic habitat they occupy is open to some very basic abiotic influences and *Daphnia* show responses to many of these. Radiation affects them both via the thermal and light regimes. All life has maximum and minimum tolerances, *Daphnia* are no exception. Most temperate species are only just active at 1 °C, with optima between 20 and 30 °C and rapid death at 40 °C. Optima vary with species. Arctic populations show tolerances 4–5 °C less than temperate conspecifics (individuals of the same species).

More subtle limits exist. Longevity is affected by water temperature. At 8 °C individuals live up to 100 days, at 18 °C to 42 days and at 28 °C only 26 days. Temperature, interacting with food availability, mediates population growth rates, highest with high food and high optimal temperature. The warmth may increase the speed and efficiency of these ectotherms' metabolism. Certainly it speeds up moulting rate, which in turn reflects faster growth. Since eventual reproduction and brood size depend on maturity, the faster *Daphnia* assimilate food, moult and grow the faster the population will expand. Egg development is faster with increasing temperature, 20 days at 5 °C, and a little over 2 days at 25 °C. Temperature controls more complex parts of the life history. *Daphnia* resist unfavourable conditions by producing resistant eggs. Temperature cues, either too high or too low, start production of such eggs. Morphological changes in individual body shape also occur, especially development of head crests and tail spines. This tends to happen cyclically and is called cyclomorphosis. The causes are several and probably all integrated to an extent, but temperature is one factor. The projections are generally credited as anti-predation devices. So, temperature itself may not be the problem but acts as a seasonal cue for when predators will be active (Fig 5.1). Migrations in the water column are another common but complex phenomenon. There may be advantages moving between water masses of different temperatures, given their ectothermic metabolism. Mobile animals show avoidance of unfavourable conditions and *Daphnia* swim away from unfavourable temperatures. In

Fig. 5.1 Typical *Daphnia* cyclomorphosis, in this instance *Daphnia retrocurva*, a common American species. The individuals without helmets or long tail spines are common in spring, the helmeted, long-tailed forms in summer. In this instance the helmets and spines confer an advantage against predation, which is heaviest in summer. Individuals without these defences have bigger broods, so are at an advantage when predation pressure is low in spring. (After Riessen, 1984)

lakes this has an additional twist. *Daphnia* show marked increases in the duration and frequency of upward swimming in response to sharp cooling. This may be a mechanism to avoid sinking below the thermocline. The temperature controls amplitude of the migration, resulting in aggregations or dispersal depending on water conditions.

Light is closely linked to these migrations. There is a general sinking by many zooplankters during daylight, or just before dawn, and a rise during darkness. This may serve many purposes but the environmental cue is light. *Daphnia* show short-term responses to light intensity and react differently to different wavelengths. Movement is predominantly vertical in blue or red light, horizontal in violet or white light. These colour dances may prompt adaptive movement as light alters over 24 hours. Light regime affects development and reproduction. The resting eggs require fertilization, whereas in good conditions reproduction is parthenogenetic. Photoperiod induces production of normally rare males.

Water body mixing and turbulence structure the distribution. The effects may be irresistible such as wind-induced currents or convectional water movements aggregating the *Daphnia*. Turbulence also induces cyclomorphosis and affects feeding. Different *Daphnia* species filter at different efficiencies. Mixing and turbulence alter food availability compared to still conditions because different foods sink at different rates. Too much mixing with excessive suspended solids can be detrimental and clear water or turbid water species exist.

Cladocera differ in acidity requirements. *Daphnia* species have been regarded as vulnerable to low pH, certainly at levels below pH 5.0, and are lost from acidified lakes. The effect may be due to a shift in competitive edge to other species, rather than the acidity itself. Laboratory tests have shown that low pH (e.g. 3.0–4.0) will kill *Daphnia*, without any competition or predation pressures. They are resistant to elevated pH and will survive up to pH 10–11. Oxygen is not normally a limitation to these lentic creatures. Larger lake *Daphnia* seem to need higher levels than small pond species (e.g. *D. pulex*), at least 1 mg l^{-1} O_2 versus 0.3–0.6 mg l^{-1} O_2, but these levels are very low anyway and most natural waters should be sufficient. At less than 3.0 mg l^{-1} O_2 haemoglobin production is stimulated.

Many other chemicals from simple ions to soluble vitamins appear to affect some aspects of *Daphnia* life histories. The major factors outlined above give a good indication of just how complex the results can be. There is one added complication. Since *Daphnia* reproduce parthenogenetically most of the time populations contain many cloned sub-populations. Although they are all the same species, in the same lake, the development of clones with different genotypes results in sometimes obviously different sub-populations, especially where migration behaviours are partly genetically controlled. Some responses blur the divide of abiotic and biotic influences. *Daphnia* tend to avoid weed beds and the shore's edges. This behaviour may be controlled by lighting or detection of plant chemical secretions. The purpose may be to avoid the sub-optimal habitat, especially lurking predators. Morphological and life history changes occur in response to the perception of chemical cues given by predators. Innate rhythms of feeding and migration have been detected in constant experimental conditions where the basic cues such as light or temperature have been removed. Abiotic and biotic factors eventually become difficult to separate.

Gammarus pulex is widespread in Western Europe. Again palaeoecology and zoogeography provide the first, big pattern. The species is found throughout mainland Britain but not, until recently, in Ireland. Introduced there, it is now spreading rapidly in an amenable environment. So the previous absence was not due to the environment, abiotic or biotic limits, but was an effect of time and space, the species not having colonized Ireland before Britain and Ireland separated. Related *Gammarus* species occur all around the Northern Hemisphere. That they are distinctly different from *G. pulex* represents a pattern on a time scale that has allowed differentiation into separate species.

Within these gross ranges in time and space *G. pulex* does not occur in all habitats or in equal abundance. Broad limits exist. It is a freshwater species and salinity, with the linked problems of osmoregulation, is a barrier to marine habitats. Equally, very barren, ionically poor waters are too dilute. The minimum for adult survival is 0.01 mM NaCl and 0.1–0.2 mM may limit juvenile survival. There are several very similar species in marine and brackish habitats and they cannot penetrate freshwaters. They may be found in inland waters where unusual ionic regimes occur, especially brackish salt pans, or where tidal influences encroach. *Gammarus* is a crustacean, a group sensitive to high acidity. The species appears restricted to waters of pH greater than 5.7 and is vulnerable to even brief pulses of pH 5.5. Oxygen is less of a constraint and *G. pulex* is tolerant of lower levels than many freshwater arthropods. Oxygen uptake changes with other factors, for example temperature, flow, substrate, salinity and individual size. Small-scale variations in oxygen availability may alter the distribution of individuals in a stream. Flow and substrate are important. Different sized individuals congregate in different sized substrate; larger individuals are found in an increasingly narrow particle size range. Larger particles provide larger interstitial spaces but this pattern is lost in static water. Note that *G. pulex* lives in both lentic and lotic waters so this effect will cause pattern differences between populations in either habitat. Particle size also affects movement, both upstream swimming and downstream drift. The movement patterns appear to show some rhythm but this may be a response to light/dark cues, not an internal clock. So as light regime changes with the seasons so will movement which is greatest in June/July and least in March. Temperature, precipitation and water level all add detail, generally promoting positive rheotaxis to current. (Note: '-taxis' is a general term for a biological movement response; 'rheo-' refers to current; 'positive' means moves towards or into, so a positive rheotaxis to current means the animal will move upstream).

Light/dark responses also affect microdistributions around sheltering rocks and in weed. High densities have been associated with weed clumps, which provide physical shelter from current (plus biotic factors such as predation and detritus accumulation for food). The weed clump populations are low in winter as floods wash out the weeds. Animals also congregate on leaf accumulations, selecting rapidly decaying leaves if available. This habitat selection reflects a biotic constraint. The *Gammarus* feed on leaf detritus and weathered, native broadleaved species provide the best food. Alternatives such as grass debris, algae and bryophytes may yield so little that animals die.

There is a point at which abiotic and biotic factors are so closely tied that a precise separation is silly. For example, the period during which females can mate is limited to just after moulting. To ensure being there when this happens, male *Gammarus* will hold and carry pre-moult females, this tandem arrangement being called precupola. There may be intense inter-male competition resulting in size selection for males best able to hold on to a female Alternatively, it has been suggested that this apparent competitive selection reflects the ability of bigger males to cope with the current, an abiotic constraint, rather than ousting smaller males, a biotic interaction. Certainly sexually active males break the general rule of size selective substrate choice, apparently roaming widely to find mates, a behavioural imperative. over-riding normal abiotic limitations. All of these abiotic limitations are further modified by the biotic interactions which are properly the subject of the next chapter.

So, even in this brief synopsis of the factors influencing the distribution of two species there are patterns from huge ranges of time and space down to daily rhythms and individual pebbles. *Daphnia* and *Gammarus* are taxonomically quite close but ecologically very dissimilar. None the less they respond to the same variety of factors. There may be a lot of different things under Heaven but what influences them are the same suite of factors. However, their responses and the relative importance of these factors are very different. This reflects their ecology: one a lotic, pelagic zooplankter; the other a benthic detritivore. Table 5.1 summarizes the hierarchy of factors and their relative importance for the two species. *Daphnia* and *Gammarus* are two well-researched genera. Such an outline of abiotic autecology could be produced for any species. Details would differ but abiotic influences operate on all species in four main ways:

1. Palaeogeographical history of a species.
2. Tolerance limits, the maxima, minima and optimal ranges for survival.
3. Alterations of behaviours and life history.
4. Alterations to the dynamics of interactions, such as competition and predation, with other species.

Table 5.1 Abiotic factors in the ecology of *Daphnia*, *Gammarus* and brown trout. Although the details of physiological tolerances, needs and responses to environmental cues are specific to the taxa, the general categories are the same for all species.

	Animal taxa		
Abiotic factors: gross categories	*Daphnia*	*Gammarus*	Trout
Palaeoecology	Very ancient genus. Many cosmopolitan species, spread with break-up of ancient land mass of Gondwanaland	European distribution of genus affected by glaciation, retreat of ice and sea-level changes, e.g. *G. pulex* did not reach Ireland	Widespread in Europe, with local races and associations with home rivers. Introduced around world in suitable environments
Physiological tolerances and responses	Temperature affects growth, breeding and morphology. Sensitive to acidity, esp. pH < 5.0. Not very sensitive to low oxygen	Salinity, oxygen and acidity affect distribution. Substrate particle size affects distribution of different sized animals. Varies between sexes	Very sensitive to oxygen and temperature. Vulnerable to acidity changes and linked metal toxins. Sensitivity of adults, young and eggs differs
Movements and abiotic controls	Vertical migrations affected by light and temperature. Passively moved by water turbulence	Drift in current. Movement patterns may alter with sex and maturity	Move to spawning grounds. Some populations migrate back and forth between sea and freshwater. Cues include flow and temperature
Sub-population differences in autecology	Parthenogenic reproduction leads to clonal populations, in the one pond. Clones may differ in responses to abiotic		Different rivers systems may support populations with home range memories that maintain discrete populations loyal to a river and adapted to local conditions
Abiotic–biotic interactions	Avoidance of weed may be due to physico-chemical cues. May be an anti-predator strategy. Anti-predator morphology changes may be induced by physico-chemical cues such as seasonal changes	Ability of different sized males to cope with different flow and substrate affects competition for mates	Basic habitat selection by fish of different age may be due to predation and competition

Given all the time and money and a heavenly host to do the necessary work it might be possible to draw up these kinds of autecological outlines for all wildlife and integrate them to unravel the mysteries of communities. This rapidly gets grotesquely complicated. Also it would not represent a complete picture. Communities are more than the sum of their parts. A step up from the individual species and its needs is necessary to look at relationships between the habititat and communities within lakes and rivers.

5.2.2 *Invertebrate communities and habitat*

It is intuitively obvious to freshwater ecologists that different microhabitats, riffles, pools, plants, substrates, within the same river or lake harbour different communities. The differences can be both qualitative (presence or absence of species) and quantitative (level of abundance varies). Habitat–community patterns have been a fruitful subject for practical purposes. Generally it is easier to maintain and manage the habitat, letting the wildlife take care of itself. As long as the habitat is conserved then the animals and plants will manage very well.

A good example is given by Jenkins *et al.*, (1984). They surveyed the invertebrate communities of the River Teifi in mid-Wales, distinguishing nine microhabitats: riffle, fast run, slow run, pool, slack, backwater, tree roots, grass roots and aquatic macrophytes (Fig 5.2). These are all habitats that can be distinguished by eye with a little practice so are eminently suitable for practical conservation work. One aim of this study was to assess the conservation value of each habitat and the rarity and uniqueness of the fauna from each. The community data were analysed to look for groupings of animals commonly found together and the different habitat types in which each group was found. Four such sub-communities of animals were identified:

1. Those characteristic of riffles and fast runs.
2. A widespread group, found in all habitats except for macrophyte clumps.
3. A group associated with depositing habitats (pools, backwaters, slacks).
4. A group common in plant clumps, roots and grass, but not in the open erosive or deposition habitats.

Evidently the full variety of invertebrates depends on the microhabitats. Loss of these, through careless engineering or flow regime changes, will knock out the communities.

Ormerod (1988) analysed invertebrate communities in another Welsh river looking for habitat associations and found differences between riffle and flat, depositing reaches and also the river margins. The analyses were taken to species level so that individual species–habitat associations can be

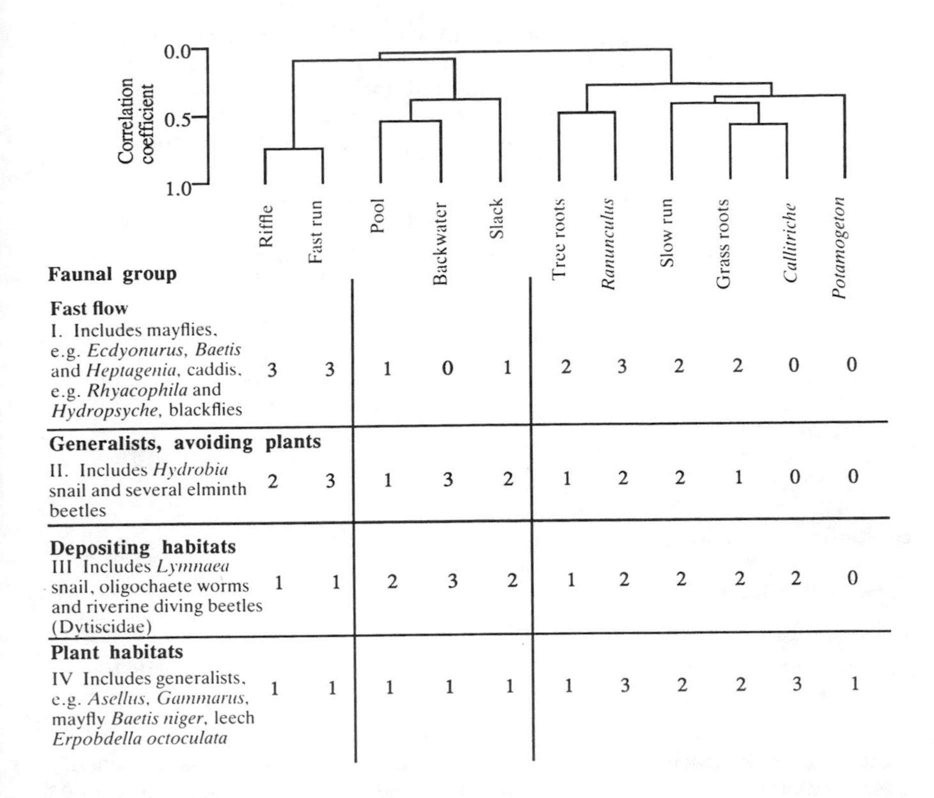

Faunal group	Riffle	Fast run	Pool	Backwater	Slack	Tree roots	*Ranunculus*	Slow run	Grass roots	*Callitriche*	*Potamogeton*
Fast flow I. Includes mayflies, e.g. *Ecdyonurus, Baetis* and *Heptagenia*, caddis, e.g. *Rhyacophila* and *Hydropsyche*, blackflies	3	3	1	0	1	2	3	2	2	0	0
Generalists, avoiding plants II. Includes *Hydrobia* snail and several elminth beetles	2	3	1	3	2	1	2	2	1	0	0
Depositing habitats III Includes *Lymnaea* snail, oligochaete worms and riverine diving beetles (Dytiscidae)	1	1	2	3	2	1	2	2	2	2	0
Plant habitats IV Includes generalists, e.g. *Asellus, Gammarus*, mayfly *Baetis niger*, leech *Erpobdella octoculata*	1	1	1	1	1	1	3	2	2	3	1

Fig. 5.2 Macroinvertebrate–habitat relationships in the River Teifi catchment. The habitats are classified by comparing the animals caught in each. The correlation coefficient is a measure of habitats' similarity from 0.0 (very dissimilar) to 1.0 (very similar). The animals made up four groups. The gross pattern of abundance of each group in each habitat is given by a numerical scale; 0=few species and usually < 10 individuals; 1 = many species but individuals typically 1–10; 2 = many species, individuals 11–100 with a few species > 100; 3 = many species with several very abundant, > 100 individuals. Numbers are for each sample period of 1 minute. (After Jenkins *et al.*, 1984)

seen. He points out that, although the correlations may be clear, the ultimate reasons why they exist are not. A microhabitat may provide the necessary physico-chemical conditions, for example flow–substrate regime, or it may contain the prey species of a predator so that the predator moves in as a behavioural response to prey but the prey may be there due to an abiotic need (Table 5.2).

Lentic systems have not been so well studied, with research concentrating on one group of animals or plant habitat–animal correlations. Different plant types support different species, abundances and diversities of animals, perhaps related to architecture or periphyton (Dvorak and Best, 1982;

Table 5.2 The importance of river channel habitats to invertebrate species. The distribution of selected species across three gross habitats of the River Wye are given as the occurrence in each habitat as a percentage of the total abundance in all three. Three habitats are distinguished: margins, riffles and flats. The differences between habitats are significant for all the species cited. (After Ormerod, 1988)

	Habitat		
Taxon	**Margins**	**Riffles**	**Flats**
Ancylus fluviatilis (river limpet)	17.1	43.0	39.8
Mayflies:			
Baetis niger	81.5	3.7	14.8
Ephemera danica	50.7	12.5	36.7
Centroptilum luteolum	83.3	0.0	16.7
Stoneflies:			
Leuctra nigra	52.9	5.9	41.2
Perla bipunctata	10.2	58.9	30.8
Caddisflies:			
Rhyacophila dorsalis	24.2	47.8	28.10

Table 5.3 Physico-chemical variables and catchment characteristics that are best predictors of stream invertebrate fauna. Macroinvertebrate communities from British rivers and a total of 28 physico-chemical, habitat and catchment factors were included in the study. The abiotic characteristics were analysed to find the set which provided the best prediction of animal communities at a site. The five best, and additional six most useful are cited (After Moss *et al.*, 1987)

Five best predictor variables	Additional six best predictor variables
Distance from source	Slope
Mean substratum size	Altitude
Total oxidized nitrogen	Air temperature range
Alkalinity	Mean air temperature
Chloride	Mean water width
	Mean water depth

Rooke, 1984; Cyr and Downing, 1988).

Our increased understanding of community–habitat ecology has vitally useful applied benefits. Much of the conservation work in rivers and streams is directed at maintaining and enhancing habitat, even where other needs, such as drainage and engineering, are simultaneously fulfilled. Maintenance of clean water and intact habitat allow the invertebrate communities to recover quickly from any short-term work undertaken and to flourish.

Aquatic habitat surveys are now regularly undertaken for conservation purposes and examples of good practice for restoration and maintenance are well documented (Newbold *et al.*, 1983, Lewis and Williams, 1984).

5.2.3 Invertebrate communities and catchments

The next step up is to consider not only the immediate habitats within a stream or lake but also the influence of the surrounding catchment and see if there are correlations between the abiotic character of the catchment and the communities. Again, work has been prompted by the increased interest in good management and the land use–water interactions. There have always been obvious examples, for instance the sparse fauna of a barren upland river or lake compared to the teeming hordes of an overgrown lowland system. Sometimes one very dominant factor has been clear, such as acidification due to catchment conditions. Teasing out the patterns in less extreme examples, and with the multitude of species involved is more difficult. The recent advances have been partly due to development of computerized techniques able to handle huge data sets of many species and abiotic factors. These analyses are able to distil the mass of data into essential patterns, picking out similarities and differences, relating these to environmental variables and integrating abiotic factors that range from the more obvious measures, such as pH or nitrate levels, to habitat features and wider landforms and uses, such as altitude, topography, type of agriculture of other land use. These analyses go by the general term of multivariate, simply because they can handle many variables of all different sorts, all at once.

In Britain this approach has been developed on a national level with the express purpose of developing a classification of running water sites. Forty-one river systems were sampled for invertebrates and twenty-eight variables, from basic abiotic factors, for example pH, oxygen through to habitat, for example macrophyte cover, substrate particle size, up to catchment factors, for example slope, altitude, stream order, latitude, longitude included. The analyses of invertebrates distinguished sixteen river types. This number is partly an artefact of the analysis used, a danger ecologists must always be on the look out for, but genuinely reflects sensible categories (Wright *et al.*, 1984). Differences arising from seasonal changes and the level of taxonomic identification used were also analysed (Furse *et al.*, 1984). The analyses picked out key indicator species which could be used, hopefully on other rivers not included in this initial survey, to fit them into the classification provided by this national data base for river monitoring and conservation purposes. Also, from the array of twenty-eight environmental variables, those which were most useful as predictors of the communities present in the rivers were picked out (Table 5.3). The purpose was primarily to identify a suite of just a few, crucial variables that could be used at new sites to predict likely fauna present and compare predictions with actual catches to look for degradation. In effect, the analyses were looking at the relative importance of the different catchment factors. Moss *et al.* (1987) carried out such an analysis. A reasonable degree of consistent prediction could be gained from just five physico-chemical variables: distance from source, mean substratum particle size, total oxidized nitrogen, alkalinity and chloride. In Britain this approach is now being developed in the water indsutry as a practical method of monitoring water quality.

A similar approach, integrating physico-chemical measures ranging from chemical parameters up to catchment land use, was applied to just two rivers by Townsend *et al.*, (1983). This allowed analysis of the community patterns down to species level and also looking at overall numbers and diversity, plus changes in trophic groups. Acidification has prompted particular efforts to look at catchment and community links (Ormerod *et al.*, 1987). These studies suggest that catchment character and water chemistry are dominant over river continuum type gradients in the communities. Certainly catchment characters set different starting points from which the continuum gradients then extend (Fig 5.3).

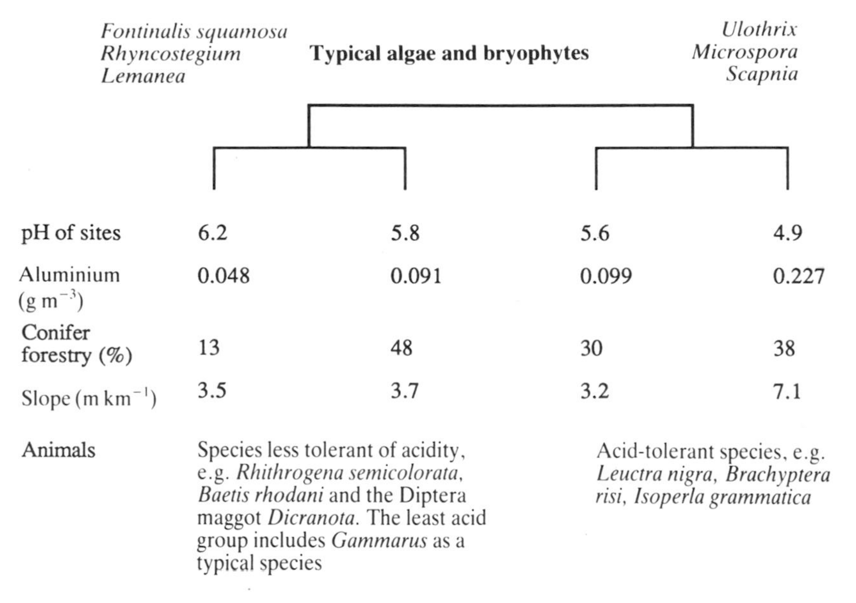

Fontinalis squamosa
Rhyncostegium
Lemanea
Typical algae and bryophytes
Ulothrix
Microspora
Scapnia
pH of sites 6.2 5.8 5.6 4.9
Aluminium (g m⁻³) 0.048 0.091 0.099 0.227
Conifer forestry (%) 13 48 30 38
Slope (m km⁻¹) 3.5 3.7 3.2 7.1
Animals
Species less tolerant of acidity, e.g. Rhithrogena semicolorata, Baetis rhodani and the Diptera maggot Dicranota. The least acid group includes Gammarus as a typical species
Acid-tolerant species, e.g. Leuctra nigra, Brachyptera risi, Isoperla grammatica

Fig. 5.3 Catchment and water characteristics and associated flora and fauna. Plant species in acidified small Welsh streams were used to classify sites into four groups. The mean pH, aluminium levels, per cent conifer forestry in catchments and slope are given for each group. Classification using the invertebrates produced generally similar groups and some typical animals are listed. The animals probably also reflect the differences in flora between sites. So animals in these animal communities reflect water, catchment and floral factors. (After Ormerod *et al.*, 1987)

Again, these fruitful approaches have not been so well worked in lentic systems. Given that algae and macrophytes reflect wider catchment influences, the invertebrates probably do so too and this remains an enticing problem. Acidification has been a major spur to studies and zooplankton communities have been linked to catchment characters but there is a paucity of work comparing communities from different lakes in relation to abiotic characteristics. Johnson and Wiederholm (1989) are a recent exception, comparing profundal macroinvertebrate communities, for many lakes and analysing the patterns in relation to abiotic features of the lakes' waters and basin characters. Depth, pH and HCO_3^-) were important abiotic factors that appeared to be linked to faunal differences between lakes.

Ponds are very sensitive to adjacent land use, but studies are limited. Barnes (1983) demonstrates how pond pH affected colonization and community development. The ponds were in old clay mining sites. Mining, with associated unusual soil mineral chemistry and drainage is often associated with acidification. The differences between ponds in this case are probably dominated by catchment inputs. Note that the immediate mode of action of the acidity will vary from species to species. For some it will be a direct autecological limit, for others, themselves tolerant of the actual acidity, a lack of plants or animals which provide shelter and food. Many pond species are tolerant of variable physico-chemical conditions. Analysis of communities in a series of small marshland pools showed different communities associated with gross flooding and drying regimes not physico-chemical variables, (Jeffries, 1989). A national pond survey to look at catchment and community patterns in British ponds has recently been set up (Pond Action, 1988).

5.3 Vertebrate communities and the abiotic environment

Links between vertebrates and their abiotic environment have been only patchily explored. Fish have been extensively studied. Amphibia and mammals have received attention largely for practical conservation. Birds have been given short shrift. There is often little integration with invertebrate work. This section concentrates on fish, firstly because of the great deal of work and secondly because this work so closely parallels that for invertebrates. Comments on other vertebrates concentrate on habitat, catchment and conservation.

5.3.1 Individual species and their environment

The autecology of many fish species is well known, with studies prompted by their economic and recreational importance. Here the autecology of one species, the brown trout, *Salmo trutta*, is outlined, in just the same way as for *Daphnia* and *Gammarus*. The hierarchy of the trout's abiotic limits are included in Table 5.1, alongside those of the two invertebrates.

The brown trout is another cosmopolitan species, polymorphic with local varieties and lives in lakes and rivers. It was originally native to Europe from Iceland to western Russia and down to Morocco. It has been introduced to all continents, except Antarctica. There is a divide between those which migrate to sea (sea trout) and those which do not. Some authorities consider that sea trout maintain distinct races with loyalty to specific coasts. Ice sheets retreated from much of its natural range some 13 000 years ago. There appear to have been two separate colonizations which, in addition to the isolation of some lake populations and river fidelity, may result in local genetic differences with consequent local variation in autecology. Just like the invertebrates there is this palaeogeographical background scale.

Temperature is a major limit to distribution. Growth occurs at 4.0–19.5 °C with lower limits for survival of 0 °C, and upper limits of 25–30 °C, depending on acclimation. Sites of successful introductions conform to these ranges. Vertebrates do not have the distinct juvenile instars of many invertebrates but developmental stages still have varying requirements. Egg development only occurs at 0–15 °C, and successful hatching is curtailed outside of 4–12 °C. Growth of the young is optimal at 13–14 °C. Acidity tolerances vary with age. Adult fish will be killed at a pH of less than 4.5, as ion balance is disrupted. At higher pH adults may survive but still be stressed. High pH is also lethal, at pH 9.0 or above. Eggs and young are vulnerable, with hatching and growth inhibited. Acidified lakes may still contain adults though young recruitment has entirely halted. The salmonids, of which the trout is a member, are sensitive to oxygen concentration. The lower limits of tolerance are about 5.0 mg l^{-1}, but generally higher levels are preferred. Oxygen depends on temperature. So salmonids tend to be associated with cool, oxygen-rich waters. Microscale differences may be important. Oxygen for the eggs depends not only on the simple amount present in the water but also on the availability in the gravel containing the eggs. A low flow of water, due to poor current or clogged substrate, can still result in oxygen starvation.

Trout populations in a river depend heavily on the availability of suitable habitat for adults, juveniles and eggs. Older trout (one year plus) prefer deeper water and cover from riparian vegetation. Younger fish stay in shallower water. This may not be a preference, but because older fish exclude them from deeper water which also contains other predators. Note how similar this is to the invertebrate examples, as habitat needs interact with biotic pressures. Trout migrate to spawn in suitable nursery areas. For lake populations this means movement into adjacent streams. Migration rhythms seem to be stimulated by river spates and water temperature. In the latter case a low threshold may be crucial. Only when temperature drops below 6–7 °C will the trout move. For sea trout the migration between salt and freshwater must involve other cues as well as a remarkable ability to adjust to the severe change in osmotic regime. Though fidelity to a particular coastline is strong, that to individual rivers is weaker. Evidence from other salmonids suggests olfactory clues are involved, perhaps the general odour of the home river, pheromones or other secretions from resident fish and this may require some imprinting on juvenile fish. For the ocean-going salmon the return to the home river may rely on electric potential generated by the Earth's magnetic field and some sense of ocean currents. Having relocated the river the actual entry is partly controlled by abiotic factors such as water temperature, light, tide and flow. Trout and other salmonid ecology is extensively described in Mills (1971, 1989a) and Elliot (1989).

5.3.2 *Fish communities and habitat*

In the discussion of trout autecology above, habitat availability was a major limitation on populations. The availability of microhabitat in a channel can shape multi-species fish communities. Gorman and Karr (1978) analyse fish community structure and compare results for a North American temperate stream and a tropical Panamanian stream. Habitat criteria were depths, corresponding to shallow edges, riffles/shallow pools, pools and deep pools, current, substrate, vegetation and miscellaneous debris such as tree trunks. The categories and measures were used to analyse and calculate indices for the diversity of habitat forms. Fish community diversities at sample stations were analysed and compared to the habitat diversity measures. For both temperate and tropical streams there was a trend of increasing community diversity with habitat diversity. Overall diversity correlated strongly with combined substrate, current and depth diversity indices. While individual components such as depth or substrate were significantly linked with species it is the combination of aquatic habitats at a site that was the best predictor of fish communities. Riparian features, such as overhangs and vegetation, should be integrated with submerged features. Accidental demonstrations of fishes' needs have been provided where human activity has destroyed or altered habitat (reviewed by Alabaster, 1985).

Channel habitat surveys for applied fisheries work have been devised, evaluating the suitability and likely crop that can be sustained. (Milner *et al.*, 1985). The habitat types identified at this level are very similar to those described with the invertebrate studies (Jenkins *et al.*, 1984). Integration of fish and invertebrate studies would be perfectly possible. Understanding the habitat needs of fish has allowed active management and modification to be undertaken to improve rivers (Alabaster, 1985). Even artificial structures that mimic natural features, such as instream features like current deflectors or overhanging shelves, provide valuable habitat (Swales and O'Hara, 1980, 1983; Brusven, *et al*, 1986; see also Fig 5.4).

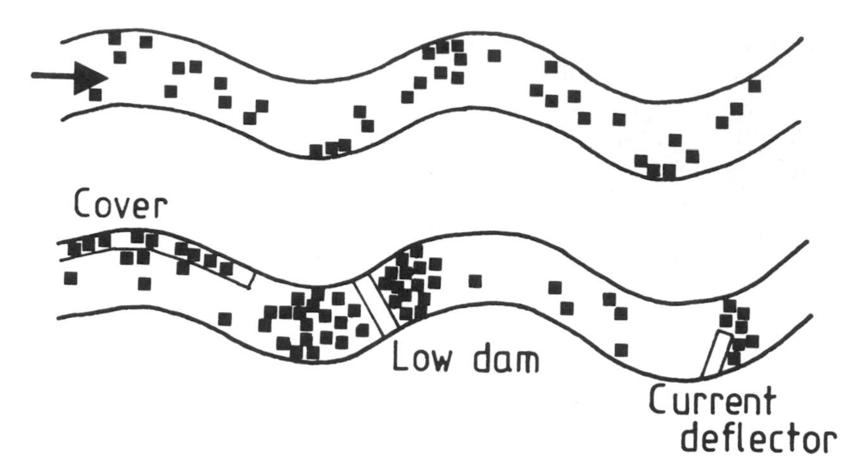

Fig. 5.4 Effect of artificial stream habitat devices on fish numbers and distribution. Instream and bankside devices were added to a stretch of the River Perry, a typical lowland, coarse fish river in Britain. Numbers and position of fish were noted by electrofishing before and after establishment of the devices. Overall numbers increase and fish are positioned by the structures. (After Swales and O'Hara 1983)

5.3.3 *Amphibians, reptiles, birds and mammals*

The variety and intensity of work on invertebrates and fish rather eclipses studies of other vertebrates, although ironically it is this latter group that attracts the bulk of attention for conservation work. The dramatic decline of animals such as amphibians and the European otter, *Lutra lutra* (L.), has inspired most of the work on their habitat requirements. Here the results of studies on amphibian habitats, riparian bird life and otters are briefly outlined. The most important lesson is that they respond in the same ways as invertebrates and fish, the same basic ecological principles are at work, although the detail differs.

One of the touchstones of threats to wildlife has been the decline of amphibian species in Britain in the twentieth century. In an attempt to unravel the causes and suggest corrective measures studies have focused on the autecology and habitat needs of the species. Cooke and Scorgie (1983) summarized trends in amphibian change throughout most of Britain. Despite the public perception of a decline the general status for most species was no change overall, with some local decreases. The rare natterjack, already patchily distributed and vulnerable to habitat destruction and disturbance, was the only one showing severe decline (Cook and Scorgie, 1983). Extensive surveys of amphibian habitats have been carried out, especially at breeding sites. Beebee (1983) was able to look at amphibian breeding sites across a soil and landscape gradient which produced different conditions in nearby pools. One newt, *Triturus helveticus* Razoumoski, was widespread, tolerating acidity down to pH 4.0–5.0 and conductivity of 64–800 $\mu S\ cm^{-1}$. Alternatively the common toad, *Bufo bufo* (L.), may be restricted to circumneutral sites. In Britain, the rare natterjack toad, *Bufo calamita* Laurenti, has been the subject of more detailed work. Temperature limits vary between adult and larval stages (maxima 41 °C and 33 °C respectively) and are greater than for the common frog, *Rana temporaria* L. (38 °C and 25 °C, adults and tadpoles respectively). The natterjacks live on dune sites, relying on the shallow dune-flashes to breed. Such sites will typically be warmer than ordinary pools used by the common frog. Temperature minima are less certain. The common frog's tadpoles will still grow at 10 °C, the natterjack's tadpoles die. Spawning will only occur at greater than 6 °C and post-hibernation emergence at 8–15 °C, compared to 4.5–10 °C for the common toad (Cooke and Scorgie, 1983). The natterjack shows autecological adaptations well suited to the habitats in which it lives, but is unable to use effectively the wider habitats used by the common frog and toad. The size and shape of ponds and the type of plant cover in and around the water are also important. All British newt species prefer ponds with some vegetation, especially genuinely aquatic macrophytes, for example starworts, *Callitriche* spp., rather than encroaching emergents. This provides cover and egg-laying sites, individual eggs being glued to the leaves. Frogs and toads like plant cover though the common frog frequently lays in shallow, temporary pools with inundated terrestrial species. Toads lay their strings of eggs around the vegetation, frogspawn floats freely. So habitat architecture affects all the amphibia at the egg-laying stage. Strijbosch (1979) presents a nice example from the Netherlands, with more species, showing preferences for different habitat architecture (Table 5.4). Surrounding land use is important too, with sheltered access to the ponds, such as dense vegetation, increasing amphibian success.

There are no native aquatic reptiles in Britain. Their absence probably is a palaeogeographical pattern, as some pond terrapins do occur quite far north on the European mainland. It would not be surprising to find they are affected in similar ways to the amphibia, especially by habitat.

Despite the intense interest in the conservation of birds and the huge recreational value of birdwatching, surprisingly little research comparable to the basic autecoloy of invertebrates and fish has been done. Some is available looking at the possible impact of acidification linked to catchment land use change, on the impact of river engineering through habitat loss and the impact of human disturbance such as boating and angling litter. The approach to studying birds and mammals differs markedly to that of other animals. The intricate interactions of physico-chemistry and communities for invertebrates have no corollary. The advent of acidification has prompted some research which suggests that birds are affected by abiotic parameters, though it may be indirectly through foodwebs. The dipper, *Cinclus cinclus* (L.), is a characteristic resident of upland rivers and streams. Ormerod *et al.*, (1986) looked at the numbers of breeding pairs on acidified

Table 5.4 Habitat selection for spawning by amphibia in the Netherlands. The general vegetation types in a series of fens were described and the proportion of each type as a % of total fen vegetation cover is given. The percentage frequency with which each vegetation type was used for spawning is shown, firstly for all nine species of amphibia in the study, then for six selected species individually. The frequency occurrence of spawning sites for all nine species (column 2) differs significantly from the frequency of vegetation types in the fens (column 1). The amphibia are selective and individual species show further specific differences in spawning site choice. (After Strijbosch, 1979)

Vegetation types	% cover of vegetation type across all fens surveyed	Frequency of use of vegetation types for spawning site by amphibia						
		By all nine species of amphibia in study	*Rana temporaria*	*Bufo bufo*	*Bufo calamita*	*Triturus cristatus*	*Triturus vulgaris*
Rooted/submerged pondweeds	1.6	19.5	13.0	5.3			
Littorella/Isoetids of oligotrophic waters	3.5	7.1	8.7		33.3		
Inundation areas, nitrophilous	1.3	0.6		5.3			
Tall grass/sedge fen by eutrophic waters	1.4	17.2	19.6	7.9	11.1	87.5	100
Small sedges, mire sites	13.8	43.8	45.7	65.9	22.2	12.5	
Sphagnum	50.7	7.7			33.3		
Willow scrub	3.6	2.4	13.0				
Open water and other types	24.1	1.8		15.8			

and unacidified streams in Wales. The numbers of likely breeding sites actually occupied were used as a measure of the impact of different factors such as pH, aluminium levels, prey abundance and land use such as forestry. There were significant decreases in the occupancy of likely sites in the acidified streams, especially where aluminium levels were high.

Habitat use along river corridors has been investigated to assess the impact of human modifications. Round and Moss (1984) present results from extensive surveys of rivers in Wales, describing habitat use by six common river birds. Links between habitat such as riffles, bedrock and shingle, plus more general channel measures such as gradient and altitude and bird numbers were analysed. Some species showed obvious positive or negative trends. For example, the moorhen, *Gallinula chloropus* (L.), commonly associated with lowland ponds and marshes, was negatively correlated with altitude and highland river feature, while the grey wagtail, *Motacilla cinerea* Tunstall and dipper show positive correlations. The greatest overall species diversity occurred in the middle reaches where upland and lowland features overlap, providing maximum habitat diversity. Just like the invertebrates, just like the fish, habitat diversity encourages wildlife diversity. During the 1970s surveys were made of bird populations along selected rivers throughout Britain (Marchant and Hyde, 1979, 1980). These data, including general classifications of river types (e.g. fast, slow, canal), allowed an overview of river preferences and changes in populations.

One of the main aims was to gauge the impact of river corridor management, pollution and disturbance. Management insensitive to habitat needs reduces species present and an abundance of those that remain (Williamson, 1971; 1980). Natural destruction such as floods reduces breeding success too, but not the overall population of breeding adults. The actual loss of habitat diversity by clumsy engineering is the problem. Equally reinstating and creating habitat encourages the wildlife. So birds do respond to the abiotic world and to habitat and often in just the same way as invertebrates and fish (Fig. 5.5).

Two amphibious mammals have been intensively studied: the otter because of its decline and the coypu, *Myocastor coypus* (Molina), because of its establishment as a pest in Britain. The European otter has declined throughout its range in Western Europe. It lives from northern Scandinavia to southern Spain, from Ireland east into Russia. Physical and chemical tolerances are wide and habitat use, be it upland rivers, lowland marshes or the sea, are catholic. Pollution, especially pesticides, as this carnivore is vulnerable at the top of the foodweb, may be partly responsible for losses. Much attention has concentrated on habitat needs. The general result is that otters can maintain a viable population, even on rivers with a considerable human presence, as long as nest sites (called holts), refuges and, above all, the general riparian cover of trees and bushes are intact. The otter is secretive rather than delicate and seems able to survive among humans as long as shelter is available, (MacDonald and Mason, 1983, 1986).

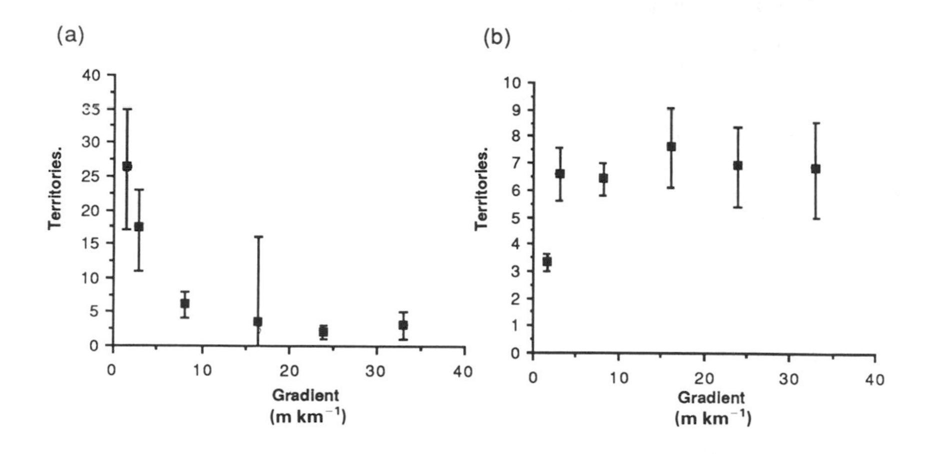

Fig. 5.5 Birds and catchment characteristics. Number of territories of (a) mallard ducks and (b) grey wagtails on rivers of different gradients. Figures are mean ± one standard deviation. Mallards are associated with shallow gradients and are common on lentic waters too. Wagtails are associated with the steep gradients of upland rivers, utilizing nest sites such as holes among boulders and cascades. (After Marchant and Hyde, 1980)

The coypu's establishment in Britain and subsequent extermination provide similar evidence of autecological and habitat factors affecting mammals. In the case of the coypu, originally from South America, East Anglia provided an ideal habitat for fur-farm escapees. A feral population rapidly established. The animals damaged farm crops and river banks in a region vulnerable to flooding. Suitable habitat allowed establishment just as habitat decline endangered the otter. An apparently successful eradication campaign has recently been completed. In part, the success has been due to modelling population trends and, in particular, evidence that coypu in Britain were vulnerable to harsh winters, a climatic limitation that allowed accurate estimates of the necessary trapping effort to be devised (Gosling, *et al.* 1988).

Research on amphibians, reptiles, birds and mamals and how they respond to the abiotic world lacks the detail and intricacy of the invertebrate and fish work. This is partly because there are far fewer species to analyse. However, even these brief glimpses show that the same general ecological pressures of autecological tolerances, habitat needs and wider catchment impacts affect all animals.

6 Animal communities and biotic interactions

The patterns in animal communities arising from the abiotic environment are complex enough. Yet simultaneously and often intricately meshed with the abiotic factors are a range of biotic interactions, colonization and migrations, competition and predation, energy flow and foodwebs. They are treated separately more from convenience and tradition but the divide between purely abiotic or biotic pressures is often unclear as shown in the last chapter where physico-chemical or habitat alterations had an effect by altering the balance of competition or predation. The same general point applies to both abiotic and biotic factors. All of them operate to some extent on all species, in all habitats but their relative importance varies. Unravelling the varying balance of factors is more useful than trying to find one single, all embracing, dominant process: it does not exist.

This chapter concentrates on experimental approaches to teasing apart the biotic processes at work in communities. Experimental manipulations of communities have assumed an increasing importance in this field of ecology. Freshwater systems are amenable to this, especially small ponds, enclosures in larger water bodies, or sections of streams from which species can be excluded or confined. Ecologists can alter the make-up of a community and follow the changes to look for the impact of competition or predation. There are two basic forms of these experiments (Bender *et al.*, 1984). 'Pulse' manipulations rely on change which may be short-lived though the consequences run on for some time after. For example, a sudden physical disturbance to a habitat which is then left to recover. A 'press' manipulation involves inflicting a change that operates over a longer period, as a continual pressure on the system. Adding predatory fish to a pond will result in increased predation pressure for as long as the experiment is continued. Allan (1984) reviews the background and practice of experimental techniques such as these in freshwaters.

6.1 Population dynamics

A population is a group of individuals of one species within a defined place, and often time. For example, the population of mallard ducks on a duck pond or the population of a species of diatom in the plankton of a lake. Often the limiting area or volume is set for the convenience of study. This results in measures of population density based on samples, which are supposed to be representative subsets of the population. Data obtained are commonly given as algae per millilitre, mayfly per square metre of substrate, watercress plants per hundred metres of river. It is not uncommon, nor terribly wicked, to allow the precise definition of a population to slip and lump species together. Ecologists often refer to the total wildfowl or fish population of a lake.

Besides these practical definitions, populations of animals and plants and how they change are the basis for so much ecological work. Some species maintain apparently steady populations from year to year; others oscillate in regular cycles; yet others fluctuate apparently chaotically. Even in apparently steady populations individuals are dying, being born, immigrating or emigrating. Populations are never static. So the study of how and why populations change is the study of population dynamics. A grasp of the basics of population dynamics is a good basis to understand other biotic interactions. If it were possible, the population dynamics of some unusually cooperative species after establishment in a habitat may begin to reveal a general pattern. There would be a steady increase in numbers, followed by a levelling out which may fluctuate a bit but is steady over the long term, barring catastrophic change in habitat or other wildlife. This general pattern arises from three integrated factors:

1. The life history pattern of the species. This encompasses the number of young produced, how quickly they grow and their longevity. Differences arise if generations overlap, for example long-lived fish species, or are discrete, for instance some insect species and whether or not individual reproduction occurs just once (e.g. the Pacific salmon, *Oncorhynchus* spp.) termed semelparity or repeatedly (e.g. *Daphnia pulex*) termed iteroparity. Even if overlapping generations occur in a population they may be obvious as different sized individuals, each wave called a cohort.
2. The eventually roughly steady level reached after the initial rise is largely a result of resources available, with competition mediating an individual's chances of gaining enough to survive, and predation, parasitism and disease taking a toll.
3. The fluctuations may be the results of life history, abiotic and biotic interactions setting up oscillations of varying degrees of regularity.

Much effort has been put into unravelling these patterns. Partly, this is their sheer fascination but there are important practical reasons such as understanding fisheries dynamics or changes in algal populations as pollution develops. There are many apochryphal calculations of how quickly we would be overwhelmed by the numbers of any particular species if it were free to produce without any constraint. This potential ability to increase numbers is common to all life, although the actual rate at which this happens would vary. Seldom do populations in the wild actually attain this rate of increase. A simple example might be a population of *Daphnia* in an aquarium. At first the population will grow rapidly, with food plentiful relative to the numbers of individuals and no checks such as competing species or predators. The population size can be measured at intervals, such as weeks, starting with the numbers at the start (week 0), then week 1, week 2 and so on. An abbreviated notation for this is N_0, N_1, N_2 and so forth.

A very general model of this population growth is given by the equation $N_t = N_0 R^t$, where N_t is the population at whatever time it is that you are interested in, for example week 4 (N_4), or week 10 (N_{10}). N_0 is the starting population and R a measure called the fundamental net reproductive rate. In the example with *Daphnia* the following very simple rules may apply: (a) each *Daphnia* gives birth to ten young a month but (b) each *Daphnia* only lives a month. The following population growth would happen, with a starting population (N_0) of ten animals.

Time	*Daphnia* at start of month	Young produced	*Daphnia* deaths during month
Month 1	10	100	10
Month 2	100	1 000	100
Month 3	1 000	10 000	1 000
Month 4	10 000	100 000	10 000

In this case $R = 10$. For example the population at end of month 1 can be worked out from the equation $N_1 = N_0 * R^1$, or, putting in the numbers observed from the real population $100 = 10 * (R)^1$. For the population at the end of month 8, $N_8 = 10 * (10)^8$, that is 1 000 000 000. R incorporates the number of young born and survival of adults through to the next generation.

This simplistic *Daphnia* example demonstrates the potential for phenomenal population growth, described as geometric or exponential. Sometimes populations will increase like this during a phytoplankton bloom in spring or the zooplankton grazers which then exploit the algae, but this rate of growth never continues for long. Growth slows and the population reaches or fluctuates around a steady state, commonly called the carrying capacity, the number of individuals of a species that the habitat can support. Technically this is that level at which birth and death rates are equal, that is there is no net change. It is a simplistic idea but has an intuitive sense. Should resources increase or decline so the carrying capacity changes and population numbers adjust (Fig. 6.1). Typically individuals will be competing for the resources, such as algae for nutrients, *Daphnia* for the algae as food, macrophytes for light, sticklebacks for nest sites. The chances

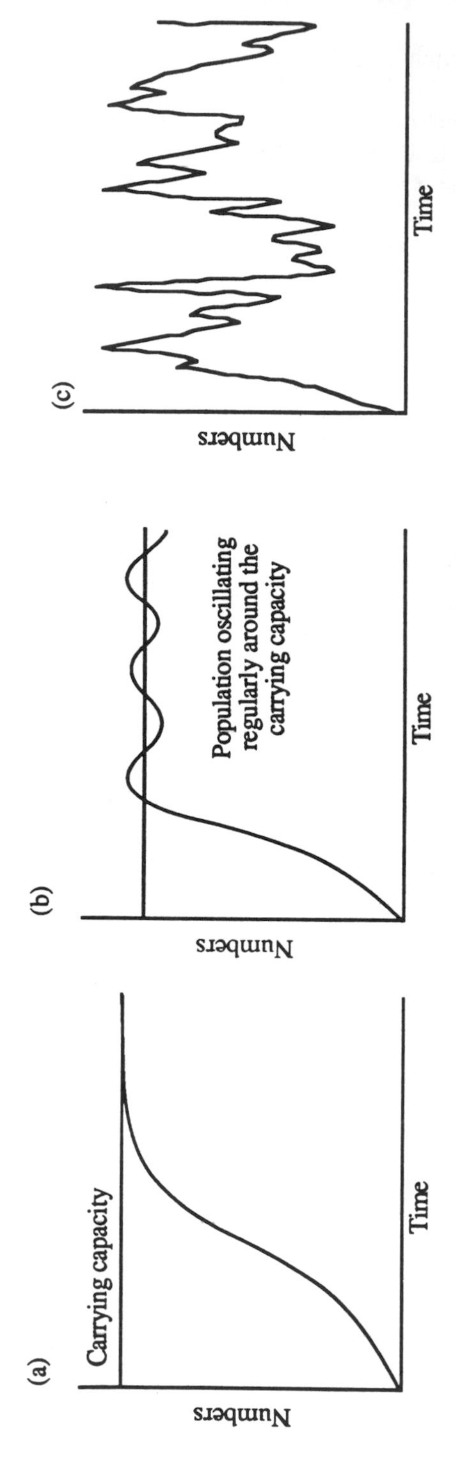

Fig. 6.1 Basic patterns of population dynamics. (a) Exponential growth reaching a steady, stable carrying capacity. (b) Exponential growth followed by regular oscillations around the carrying capacity. (c) Growth followed by apparently chaotic dynamics.

of any one individual surviving and thriving depend on the overall population of compatriots. Where there are few individuals, and assuming other pressures such as predation are not a problem, survival will be good. As the population rises the chance of any one individual surviving decreases, as resources become scarcer and competition increases. The likelihood of individual mortality increases as a result of the increasing numbers of others in the population. This is called density-dependent mortality; the individual's chances of survival are altered in direct proportion to the total numbers in the population. So the carrying capacity might be determined by density-dependent mortality, though in reality this can be uncommon as other processes take their toll. Some changes might be density-independent, that is their impact on an individual does not depend on how many other individuals there are in the population. Should a pond dry out completely it will kill all the *Daphnia*. It does not matter whether there are 100, or 1 000 000, the individual's chance of survival is still zero as lethal tolerances are exceeded. An individual's survival does not vary as a result of the numbers present. A real example is given by George and Edwards (1974), analysing population dynamics of *Daphnia hyalina* Leydig in a reservoir over two years. The changes in population were monitored. In spring there was a rapid exponential increase when resources were abundant. The rise then slowed and summer populations oscillated around a carrying capacity attributed to density-dependent factors. This was followed by a population crash in autumn (Fig. 6.2). Simple mathematics were used to work out the instantaneous population change, birth and death rates.

To summarize this basic introduction to population dynamics, populations have the potential to grow rapidly, typically reaching an approximately stable density at a carrying capacity. Populations change as density-dependent and density-independent pressures interact with life histories and reproduction. Just as with the abiotic factors discussed in the last chapter quite what happens in real multi-species communities is another matter.

6.2 Competition in freshwater communities

In the preceeding section on basic population dynamics there is a very strong theme of competition between individuals of a species and resulting population regulation by density-dependent mortality. Competition between individuals of the same species is termed intraspecific. That between individuals of different species is interspecific and may have a very important role in deciding which species live where and in what numbers.

It is useful to distinguish two ways of competing. One, generally called interference, involves direct interactions between the animals involved, to the detriment of one or other, or perhaps both. A snail grazing over a stone will bulldoze the tubes of midge larvae in its path, interfering with their

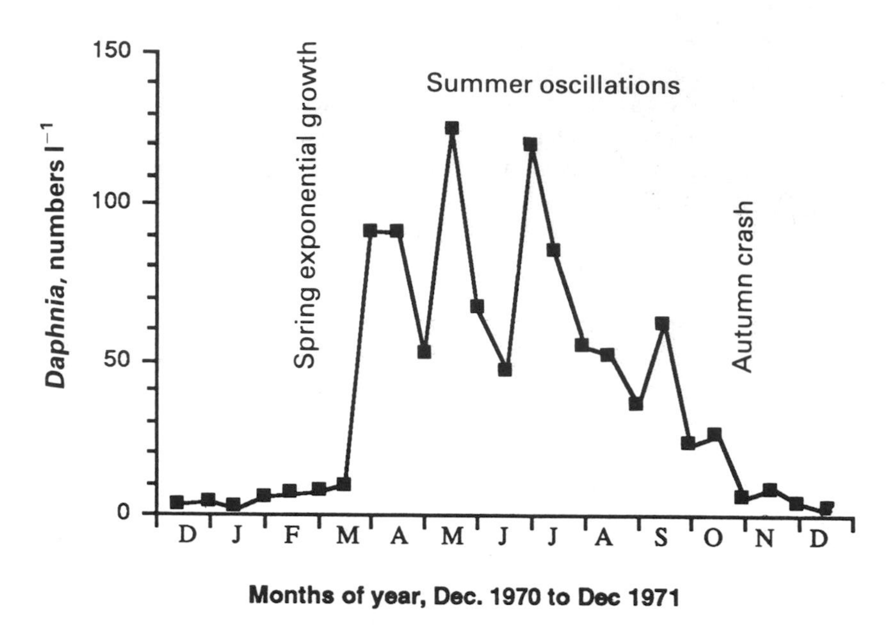

Fig. 6.2 Dynamics of a real population. Numbers of *Daphnia hyalina* in a reservoir over the course of one year. Data are mean *Daphnia* per litre. (After George and Edwards, 1974)

feeding. The alternative is exploitation, or scramble, competition. The competitors are each trying to gain as much of the resource as possible and so affect each other by the impact each has on how much of the resource is left. The competition takes its toll through the resource but there is no direct effect such as fighting or poisoning. Different *Daphnia* species filter at different efficiencies. The more efficient species will exploit more of the algal food available, its population increasing proportionately faster. Eventually the less able species may die out altogether, but individually animals never come into conflict.

The central point is that the balance of competition between winners and losers may be responsible for determining how many species can coexist in a habitat. It is worth noting that taxonomic affinity, how closely related species are, is not necessarily important. What matters is what they do and what resources they use. Very distantly related creatures can compete intensively for the same resource. In the Andes, fish and flamingo compete. Both filter plankton and benthic invertebrates from the lake water. Flamingos avoid lakes with high fish populations, but will feed in areas previously avoided if fish are excluded. The precise importance of competition has been the source of heated debate in ecology. As with all ecological processes there are circumstances in which it is predominant, others in which it is trivial. Evidence for the importance is particularly clear from manipulation experiments, of which the fish exclusion in the Andes is an example, where one or more competitors is added or excluded from a system and the resulting changes in the rest of the community can be followed and compared to unmanipulated, control communities.

A typical example which raises several subtle points is given by Cuker (1983). Cuker looked at competition and coexistence between the animals living on rocks in the littoral of an Arctic lake. The community consisted of a snail (a *Lymnaea* species), which grazes algae and chironomid midge larvae, either free-living or tube-dwelling, both types grazing algae, and free-living predatory species. Note that the herbivores all make their living in the same way, grazing the epilithic algae off the rocks' surface. Such a group of animals exploiting the same resource makes up a guild, named after the medieval guilds of craftsmen, such as barrel makers or wheelwrights who all made their livings doing the same job. Algal encrusted rocks were placed in chambers under water, then different densities of snails were added to different chambers: 0, 1, 3, 10 and 25. There were four enclosures of each of these five treatments. This therefore is a press manipulation, with the snails present throughout the duration of the experiment. The chambers were left in the lake for just over a month in summer, then retrieved and the populations of animals in each counted. Several different aspects were analysed:

1. The biomass of epilithic algae in relation to snail, to see if snail density is affecting the resource.
2. Snail growth in relation to snail density, to look for intraspecific effects.
3. Changes in other species' densities in relation to snail density (interspecific effects) (Fig. 6.3).

The dry weight of epilithon fell with increasing snail density, suggesting snails depress the food resource. Individual snail growth also varied with density. At 1 or 3 snails per enclosure individuals grew over the course of the experiment, at densities of 10 or 25 individuals lost weight. So resource depression and intraspecific competition occur, suggesting interspecific effects may also show. With just 1 snail present the midge guild hardly differed at all from the no snails treatment. At all other snail densities free-living and tube-dwelling midge numbers declined. The tube dwellers were especially affected with biomass losses of up to 90 per cent. The competitive effects may work in many ways. The snails' progress over the rocks will destroy tubes, perhaps the occupant too, resulting in costs to surviving midge larvae of time and energy rebuilding. This is a good example of interference type competition. Free-living midges will have to move out of the way, again wasting time. The snail grazing may alter the periphyton itself, changes in quantity and quality affecting midge development. Interestingly, at the lowest snail density the free-living midges may have fared slightly better as tube-dwelling species were still badly disrupted, releasing the free-

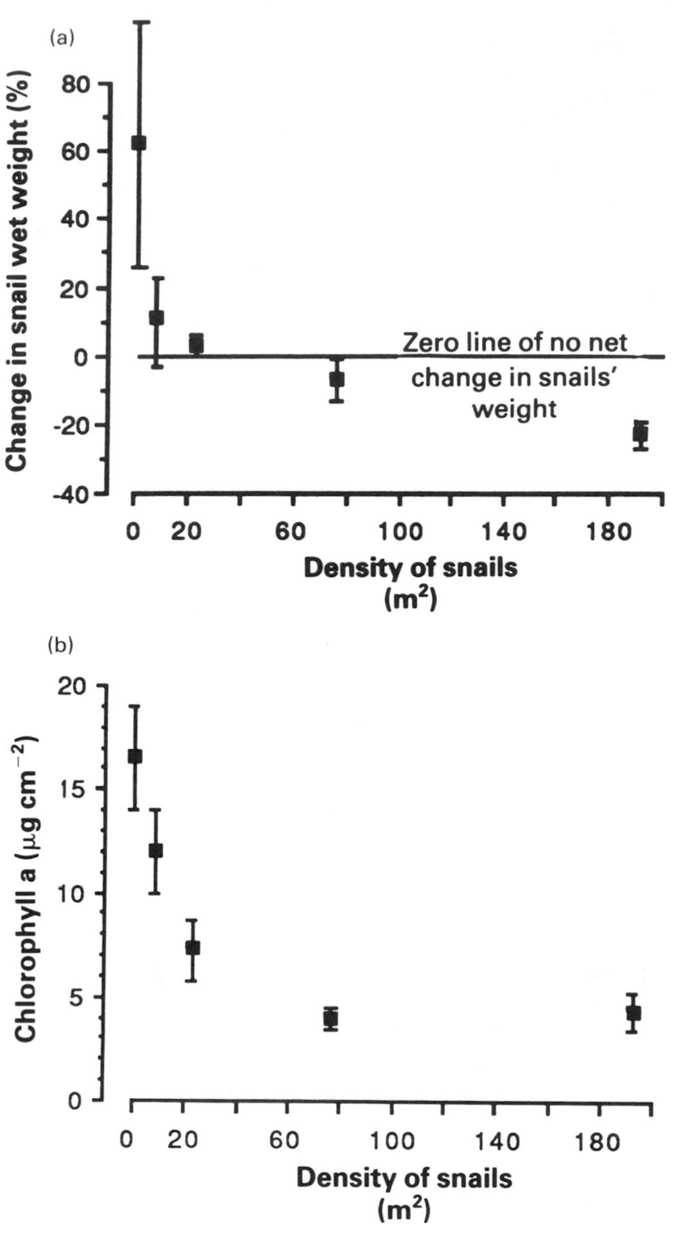

Fig. 6.3 Competition in freshwaters. (a) Percentage change in individual snail wet mass of *Lymnaea* snails added to enclosures at various densities. Net losses of weight occur at the higher densities suggesting intraspecific competition is strong. (b) Impact of increasing snail densities on periphyton (measured by quantity of chlorophyll a, $\mu g\ cm^{-2}$). Snails markedly depress the quantity of algae. This reduction in the food resource will not only affect the snails but also other competing species. Data are means ± one standard error. (After Cuker, 1983)

living midges from this additional competition while snail pressure was still not too heavy.

Such an approach has also been used in streams. McAuliffe (1984) uses exclosure techniques, in contrast to the enclosures used by Cuker. *Glossosoma* (Trichoptera) larvae were excluded, or at least greatly reduced in number, by some bricks placed in a stream using a barrier of Vaseline which they could not climb across. Periphyton and mayfly grazer numbers increased on these protected bricks. Unlike the snail example there is no obvious interference, the tough jaws of *Glossosoma* may reduce the periphyton turf so much that the mayflies cannot glean a living, an example of exploitation competition (Fig. 6.4).

The precise value of manipulation experiments is still argued (Miller, 1986; Hart, 1986b) but certainly in many grazing communities with a basic food resource on an architecturally simple surface competition appears to be very powerful.

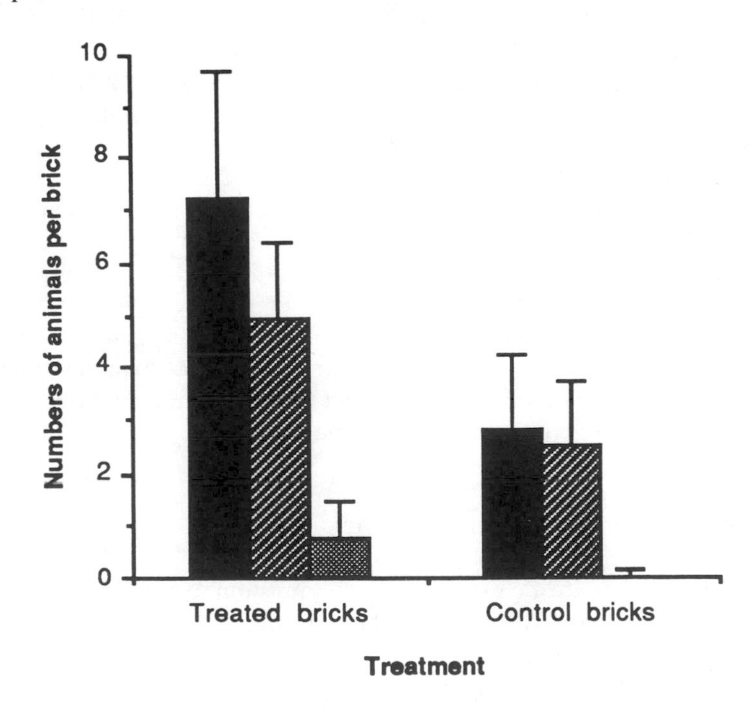

Fig. 6.4 Interspecific competition among stream grazers. Larvae of the caddis *Glossosoma* were excluded from bricks placed in a stream by a Vaseline barrier. The numbers of other grazers colonizing the bricks, by swimming or settling from the current (which the *Glossosoma* could not do), were compared to the numbers on control bricks lacking the barriers. Numbers of total grazers ■, *Baetis* ▨ and *Ephemeralla* ▩ mayflies on treatment and control bricks are shown. Data are means ± standard deviation. Numbers of these other grazers are higher where the dominant *Glossosoma* competitor is excluded. (After McAuliffe, 1984)

One strategy creatures may adopt as a response to competition is to defend a patch, a territory, of resources. Again stream communities have yielded several examples. Filter feeders, often fixed to one site and constructing nets to catch food (e.g. some caseless caddis) are aggressive to neighbours, intra- and interspecifically. Hildrew and Townsend (1980) describe interactions between net-spinning *Plectronema* in which intruders and residents fight over net sites. Even seemingly less aggressive species can still contest. Hart (1986a) describes blackfly larvae behaviour suggesting strong intraspecific interactions between these filterers. Larvae will jab at neighbours upstream (and only upstream) to try and move them out of the way, presumably so the upstream rival does not filter out the food before it reaches the downstream animal. Such aggression declines if food is increased. The result is that adjacent larvae space themselves out regularly. Damselfly nymphs also defend hunting perches and engage in displays and physical attacks on rivals, for example larvae of the common red damselfly, *Pyrrhosoma nymphula* (Sulzer) (Harvey and Corbet, 1986). Competition between adults in damselflies and dragonflies is also intense, in this instance to hold breeding territories and win mates (Corbet, 1962), the outcome of this terrestrial phase perhaps then altering the aquatic populations.

To sum up, competition, intra- and interspecific, can be a powerful biotic force in freshwaters. It will affect life histories and success of individuals. Short-term, small-scale competition, such as that by the grazers, may reduce or completely exclude inferior competitors. The diversity is lowered. In the longer term this will act as an evolutionary pressure as species adapt to avoid competition, dividing the world up between themselves. Communities as we see them today may reflect past competition, which has promoted diversification in the long term, so that actual competition is not so intense now. In many cases the communities we see reflect the 'ghost of competition past', as much as any competition present (Connell, 1980).

6.3 Predation in freshwater communities

Predation is a biotic process very like competition. Sometimes it can be dominant, sometimes trivial. It can be a force promoting or destroying diversity. Competition and predation are often so intimately tangled, their results so similar, that teasing the two apart is difficult. They have become two terrible twins in ecology and their relative importance the source of acrimonious debate.

Firstly, it helps to realize what predation encompasses. Although it may seem very obvious a definition that embraces all the ecological detail of predation, without becoming so all-embracing as to be meaningless is difficult. There are plenty of obvious examples of freshwater predators and their prey: a trout eating a stickleback; a kingfisher a minnow; a waterbeetle a tadpole; a *Hydra* a *Daphnia*. These are all variations on a theme of one animal capturing and eating another individual. Size may blur an appreciation of the process. A large planktivorous fish straining tiny zooplankton from the water is still a predator. There are many forms of parasitism in freshwater. Parasitism involves feeding on a host without attempting directly to kill it, although the damage may prove fatal none the less, for example leeches and water mites that suck the body fluids of prey, lampreys (*Lampetra fluviatilis* (L.), that rasp out tissue from the flanks of other fish. Certain insect parasites that live inside their hosts, from which they eventually emerge killing their victim, are different again. Their host is alive until the parasite emerges. They are termed parasitoids. Disease organisms such as bacteria and fungi are parasites. In the popular imagination there is a distinct difference between predation and herbivory. This is less clear in ecology, since both involve one living thing eating other living things. Taylor (1984) gives a pragmatic definition that predation occurs when an individual eats other living things. Keep this in mind, but this section will deal specifically with animals eating other animals.

It is worth noting at the outset that predation has much the same kind of impact as competition. It affects life histories and survival plus the diversity of communities, differing in short-term and long-term (evolutionary scale) results. Again much of its effect may now be evident as the result of adaptation in response to past predation: the 'ghost of predation past' (Fig. 6.5).

Manipulation experiments have been a fruitful approach here too. Murdoch *et al.* (1984) describe the results of the experimental addition of the predatory bug *Notonecta hoffmanii* Hungerford into small cattle troughs. This is an enclosure, press experiment, essentially just like Cuker's design to look at competiton. The troughs contained simple communities of zooplankton and juvenile mosquitoes. Each trough was divided in half by a partition and *Notonecta* added to one side, and removed from the other if present naturally, so that the effects on the communities could be analysed by comparing the results for each half of the same trough. The prey populations were sampled over several years. Mosquito larvae and *Daphnia pulex* were severely reduced, frequently exterminated, when *Notonecta* were added. Smaller, less vulnerable zooplankters were reduced though less severely. These troughs were small and very simple, lacking any plant cover or great variety of prey, but they provide good evidence of the destructive potential of predators. In more complex, natural systems predation may selectively remove a few most vulnerable species., The selection may be an active choice by the predators many of which, vertebrate and invertebrate, have sophisticated behaviours to maximize efficiency of prey capture. These behaviours, such as prey choice or searching out and staying in prey aggregations, are termed optimal foraging. Equally, some characteristic of the prey that makes them especially vulnerable, even to unsophisticated predators, may result in predation falling heavily on just one species. Selective predation often falls on particularly abundant prey. These may

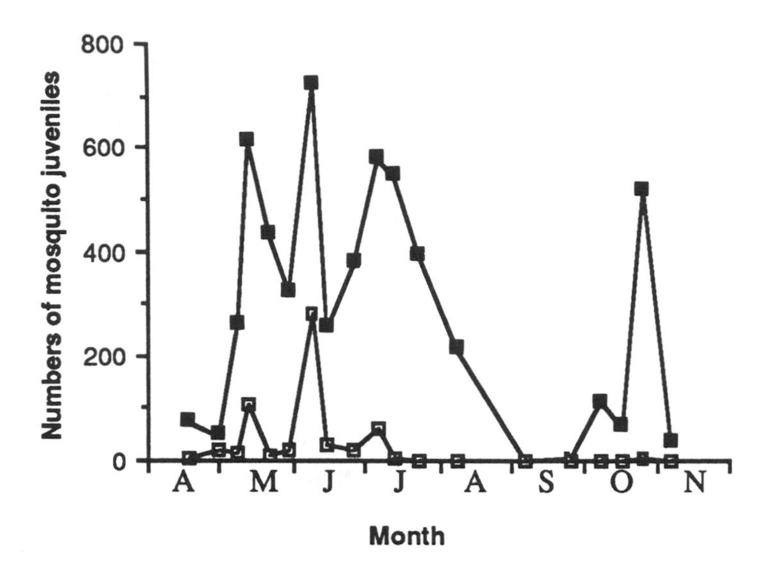

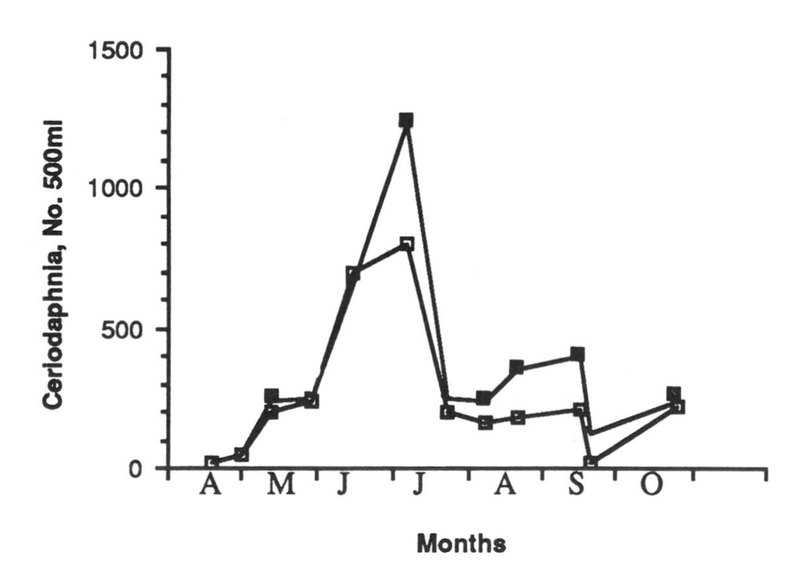

Fig 6.5 Impact of predation in freshwaters. (a) Numbers of mosquito larvae and pupae in a cattle trough, divided into two halves, with waterboatman predators added to one side (□), the other side a control with no added predators (■). (b) Numbers of *Ceriodaphnia* in same trough. Losses of mosquitoes are heavy, they are very vulnerable to the predatory bugs. Losses of *Ceriodaphnia* increase once mosquito numbers have been reduced. The *Ceriodaphnia* are much more numerous than the mosquitoes and the bugs do not inflict such catastrophic losses on the waterfleas. (After Murdoch *et al.*, 1984)

well be the best competitors, hence their abundance. By predating these, the predators can stop a few species monopolizing the resources available and this balance of competition and predation maintains diversity. Predators acting in this role have been termed 'keystone' predators (Paine, 1966). Their presence or absence has a vital role in deciding the form the community takes.

Enclosures have also been used in streams and a cunning example is described by Peckarsky and Dodson (1980), using large stonefly larvae as predators. Cages were placed in streams and one of four different treatments assigned to each cage. Firstly, no additional animals, so the effects of the cage itself could be detected. The other three treatments all involved enclosing animals in the cage, either a large predatory stonefly, free to roam around all the cage, or the larvae of the same species but restricted inside an inner cell within the main cage, or a large herbivorous stonefly, free to roam all the cage. Cage mesh sizes prevented stonefly escape but were large enough to allow likely prey, especially mayflies, to move back and forth. Numbers of mayflies were reduced in the presence of the roaming predatory stoneflies, perhaps due to the prey they consumed but also by avoidance reactions on the part of the prey responding to tactile and non-tactile cues. The restricted stonefly treatments also resulted in prey reductions, in the cage outside of the inner cell. Since the predator could not be eating the prey outside of the cell the reductions suggest that avoidance reactions by prey are part of the cause. Numbers of prey colonizing the cages with the herbivore stonefly were not markedly reduced, though there may have been some subsequent emigration. These results suggest that not only do invertebrate predators have a noticeable impact in lotic systems but also that the interactions are subtle with prey able to detect and react to enemies (Fig. 6.6).

The precise importance of invertebrate predation remains uncertain. Peckarsky (1984), in an excellent review of freshwater invertebrate predator–prey interactions, suggests this is partly due to a paucity of reliable data, though evidence is increasing. Evidence for the impact of fish predation, on both planktonic and benthic communities, is much more coherent.

Firstly an example, showing that the notonectid predators used by Murdoch *et al.*, in the cattle troughs are themselves vulnerable. Bennett and Streams (1986) looked at the distribution of *Notonecta* species in ponds with or without predaceous fish. Fish severely reduce the abundance of Notonectidae, although the diversity of species was higher in fish ponds. The most abundant species when fish were present was strongly associated with vegetation cover. When fish were absent the dominant species was one that ventured into open water (Table 6.1). This is a simple system but points up some basic principles very well; fish can exert a severe influence, altering the abundance and variety of species and cover is an important factor. Healy (1984) reviews the general impact of fish predation, detailing the

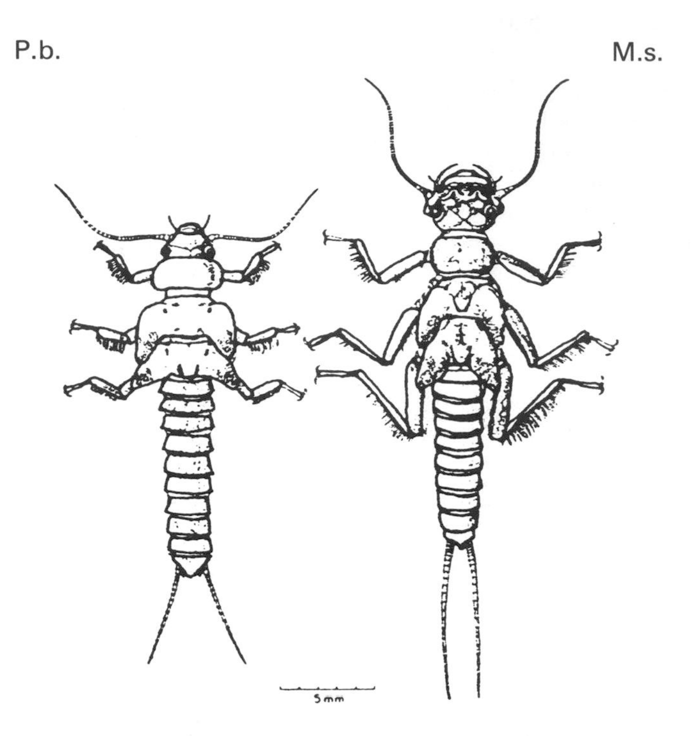

Fig 6.6 Stoneflies used in stream manipulation experiments to look at the effect of predators on prey distributions (Peckarsky and Dodson, 1980). *Megarcys signata* is a predator, *Pteronarcella badia* a herbivore. The predatory species reduced numbers of prey inside cage enclosures. The similarly sized herbivore had only a weak negative effect on 'prey' numbers, suggesting that the reduction in prey is not simply due to a large animal moving about and causing disturbance, but is due to prey eaten and prey leaving in response to danger.

Table 6.1 Mean numbers of *Notonecta* species in samples from two ponds, with or without fish, in early and late summer. The overall *Notonecta* numbers in samples from the fish pond are less than the fishless pond but the fishless pond supports fewer species, with one species numerically dominant. Data are mean numbers per sample ± one standard error (After Bennett and Streams, 1986)

	Barrows pond (fish present)			**Campus pond (no fish)**	
Date	*N. irrorata*	*N. lunata*	*N. undulata*	*N. undulata*	*N. insulata*
Early summer	0.20 ±0.10	0.96 ± 0.37	0.56 ± 0.21	1.52 ± 0.52	0.41 ± 0.18
Late	0.03 ±0.03	0.40 ± 0.14	0.37 ± 0.14	4.87 ± 1.52	0.10 ± 0.07

complications and contradictions that can arise. The impact of fish on zooplankton has been particularly heavily researched. Zaret (1980) summarized much of the work. Generally fish, attacking zooplankton in the pelagic zone, inflict severe losses on the larger species, resulting in communities of smaller, less visible species. Introduction of fish into lakes for recreational and fisheries purposes can have a profound impact on the other animals.

Just as competition gives rise to complex behaviours so does predation. Two major aspects are prey selection and foraging behaviour by the predator and behavioural adjustments by the prey in response to a perceived threat. Predators, from invertebrates to mammals, show flexible hunting behaviours to optimize their success. Essentially the behaviours are intended to maximize the returns (i.e. prey caught, nutrients gained) for the minimum effort and risk. Many predators adjust hunting in response to prey numbers. When prey are few, hunting effort may be increased, for a period, but if returns remain poor hunting is curtailed. Behavioural changes help hunters to find good patches of prey then stay in those patches, until such time as they have exhausted the prey supply. The relative abundance of prey types may cause predators to switch their efforts towards just a few, or only one, prey type. Typically, this means catching disproportionately more of one prey species than its abundance in the habitat, relative to other prey.

As the predators alter their behaviour so can the prey alter theirs to avoid capture. Movement patterns and the use of refuges alter in the presence of enemies. The senses of prey may be subtle enough to distinguish predators from among other innocuous but apparently similar stimuli. Larvae of some damselflies become much more inactive in the presence of predators (Heads, 1985) and hunting success may decrease (Jeffries, 1990a). Their response is not simply a general nervousness to any disturbance. The inactivity is much greater in the presence of genuine danger, for example fish or notonectid bugs, than with innocuous disturbances, for example bubbles, or non-predatory species which are superficially very similar in size and behaviour to the predators such as large herbivorous corixid bugs (Heads, 1985). In the example of a stream manipulation cited earlier, the prey appear to distinguish large predatory and large herbivorous stonefly larvae.

Over the longer term, predation may encourage diversity much as competition does, at an evolutionary scale. Selective pressure to defeat and baffle enemies is intense. Prey will develop better and different defences. The need is not necessarily to be completely invulnerable to a predator. A defence need not be totally effective, it need only be relatively more effective than that of other prey species, so that predation falls on them. Prey are in effect competing to develop better defences than each other. Whatever form these take, size, shape, speed, toxins, camouflage, shells, microhabitat, they all represent what has been termed 'enemy-free space' (Jeffries and Lawton, 1984).

Competition and predation remain two tangled and mysterious topics, with a lot more to be found out about them. Excellent reviews can be found in Peckarsky (1984), Kerfoot and Sih (1987) and Hildrew and Townsend (1987).

6.4 Herbivory and primary consumption

The dividing line of herbivory and predation is blurred as the preceding section points out. If this is so then the patterns and processes in herbivore–plant systems ought to show many of the general patterns that arise in predator–prey systems. The herbivores impose the same pressures as predators, the plants respond as prey, though responses may be less obvious than the escape reactions or habitat shifts of prey animals.

The impact of herbivores tends to be looked at in three areas: herbivores and phytoplankton; herbivores and periphyton; and herbivores and detritus. The latter might be described as detritivores, but it is the microflora living on the detritus that provides much of the nutritional content. Herbivory on the living macrophytes, which is such a dominant factor in terrestrial systems, has received only scant attention, essentially because it hardly seems to happen. This may represent a major difference between terrestrial and aquatic systems. To point up the recurrent ecological themes it is worth reiterating some of the questions and answers raised by predation to see if these hold true for herbivore–plant interactions. Do herbivores inflict significant losses on plant populations? Are diversity, biomass and community patterns altered by herbivory? Do herbivores show complex behaviours? Do the plants react to these attacks?

Cuker's work on competition described earlier also showed the impact that snail grazing had on periphyton. Manipulation experiments have been run to look specifically at herbivore–plant interactions. Lamberti and Resh (1983) present results from an exclusion manipulation, where grazing caddis larvae, of *Heliocopyche borealis*, (Hagen) were excluded from tiles and algal crops measured in comparison with periphyton on tiles exposed to the grazers. Ungrazed swards had higher quantities of algae, the standing crop, but the grazed swards had a higher algal productivity. This means at any one time there were more algae on the ungrazed tiles, but more were produced on the grazed tiles with this greater production being cropped by the caddis. A measure of oxygen production, which reflects live algal activity, showed that more oxygen was produced per unit of algal biomass on the grazed tiles. So, although there were fewer algae present at any instant on the grazed tiles the community appeared to be healthier and more productive. The taxa present altered too. Grazed swards were dominated by thin films of diatoms, ungrazed by thick mats of filamentous green algae. The herbivores acted just like keystone predators, their presence or absence deciding which form the plant community would take. Similar effects are reviewed by Lamberti and Moore (1984) from a wide range of other freshwater studies. So grazing does affect productivity, turnover and diversity, just as predation does, and just as grazing does in terrestrial systems.

Lamberti and Resh's experiments also analysed the behaviour of the grazing caddis. Larvae were aggregated on swards of a higher standing crop to start with but as grazing reduced the algae the caddis became randomly scattered. The herbivores were showing similar foraging behaviour changes as predators, concentrating on rich patches of food until such time as these were exhausted. This example may be simply in response to total quantity of algae but herbivores can be selective of plant species and quality. Lodge (1985) demonstrates macrophyte substrate specificity by different species of gastropod snails which appear to represent a selection for different epiphytic communities on different macrophytes (Fig 6.7). Macrophytes may suffer if smothered by algae. Bronmark (1985) presents evidence that shoots of hornwort, *Ceratophyllum demersum,* attract grazing molluscs, by release of carbon compounds that are not simply products of decay, and that the grazed stems grow better, measured as both length and weight, than ungrazed stems (Fig. 6.8) which remain coated in epiphyton.

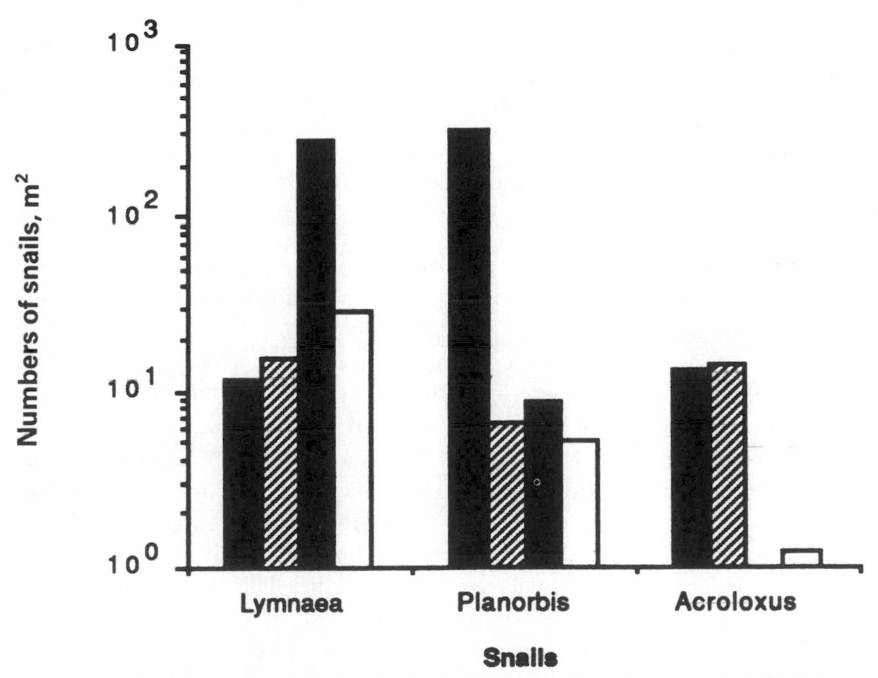

Fig 6.7 Macrophyte substrate choice by the snail species: *Lymnaea peregra, Planorbis vortex* and *Acroloxus lacustris.* The plot shows the numbers of each species on four substrates: ■, submerged parts of emergent species; ▨, lilies ■, submerged species; □, detritus. The preferences shown by *L. peregra* and *P. vortex* are linked to epiphytic communities on the different substrates. *A. lacustris* may be limited to substrates offering simple, flat surfaces due to its large, flat 'foot'. Data are grand means of numbers of snails per sample from several months samples. (After Lodge, 1985)

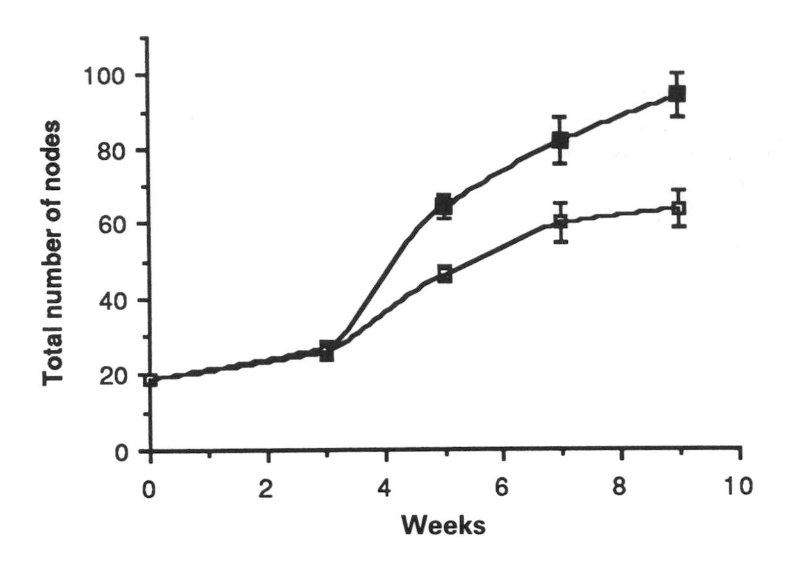

Fig 6.8 The effect of grazing by snails on the growth of strands of the submerged plant *Ceratophyllum demersum*. Growth of grazed (■) and ungrazed (□) strands was compared by counting the number of nodes along stems. Growth of the grazed stems is significantly greater after 5 weeks, as the snail grazing removes smothering epiphytic algae from the macrophytes. Data are given as means and standard error. (After Bronmark, 1985)

The zooplankton that feed on phytoplankton are also grazers. The only real difference in their interactions lies in the mechanism of capture. Effects on the abundance and diversity of phytoplankton, and the herbivores' behaviour, are similar to those of periphyton grazers. Bailey-Watts (1986) provides a nice example based on phytoplankton and zooplankton in Loch Leven, a shallow, eutrophic loch in Scotland. Through 1979–82 the phytoplankton community was sampled, taxa identified and also measured, using the longest axis of the cells. Cell size may affect vulnerability to filter feeding zooplankton, due to the mesh size of their filtering apparatus. Early in the season small algae (less than 15 μm) dominated but later in the year larger species became relatively more abundant. The decline in smaller algae coincided with the increase of *Daphnia* in all four years. The zooplankton appear to take a significant toll of the phytoplankton and there is a size-selective bias. Zooplankton grazing is indeed very like any other grazing. The herbivores are able to cause significant changes in phytoplankton communities with foraging behaviours and preferences adding to the complexity of the results (Fig. 6.9).

Detritivore-based foodwebs are very important in many freshwater systems. Petersen and Cummins (1974) presented a summarized budget of the effectiveness of detritivore activity at processing leaf litter. They

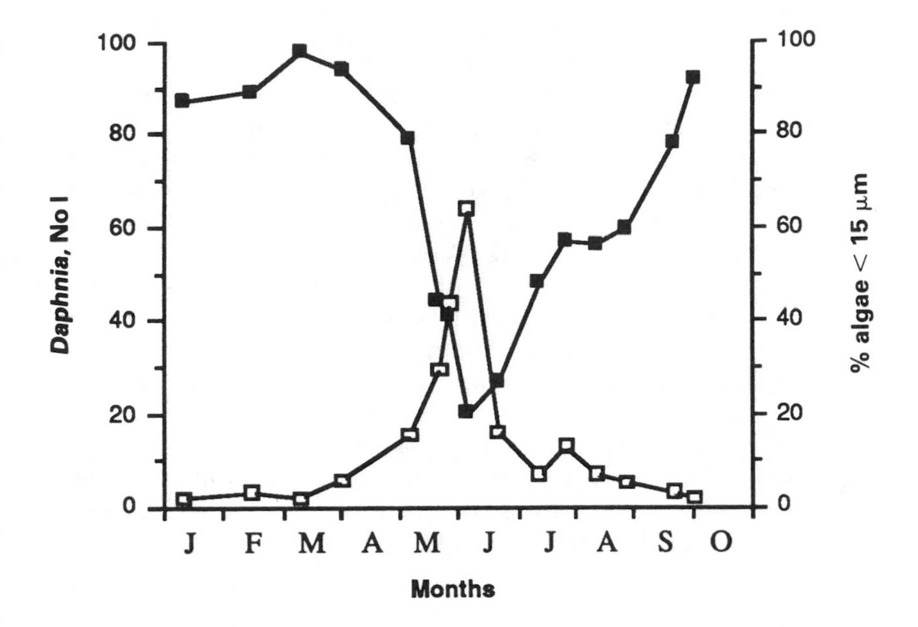

Fig. 6.9 The impact of zooplankton grazing on phytoplankton populations. The numbers of *Daphnia hyalina* in Loch Leven, Scotland, during 1982 are shown, along with the proportion of algae, in the total phytoplankton population, of less than 15 μm in maximum length. These smaller algae are vulnerable to the *Daphnia* filterers and changes in size class proportions in the algal population appear to reflect the impact of the grazers' activity. (After Bailey-Watts, 1986)
■ = Phytoplankton □ = Daphnia

analysed the breakdown of leaf packs in a stream differentiating between leaching, microbial and animal activity. Some 60 per cent of a leaf would be processed in 120 days, and after a year almost the whole leaf would have been cycled back into the system, including debris washed further downstream. Once again detritivore–detritus interactions are as complex as any other grazing. The type (e.g. species), age and condition of substrate greatly affect the colonization by detritivores and value of the food source. Some leaves make better substrate than others. This becomes particularly important when considering riparian and catchment plantations that may introduce poor quality food, such as exotic species or conifers, into the habitat.

There is some grazing damage to living macrophytes, but the extent and number of species involved is nothing like the contest between plants and their herbivores on land. Few insect-macrophyte grazers exist under water, while the majority of terrestrial species of animal life consist of the insect herbivores and their predators. The larvae of a few moths and caddis attack macrophytes, though recent evidence suggests grazing damage to macrophytes might be greater than is commonly supposed (Sand-Jensen and Madsen, 1989). Emergent plants are vulnerable and research has been done on the use of grazers to control nuisance plants, often floating species such as water fern or water hyacinth, for example control of the floating fern, *Salvinia* (Room *et al.*, 1981) but also submerged plants, in particular using the grass carp, *Ctenopharyngodon idella* L. Quite why the lush aquatic plants are under-utilized is baffling (Table 6.2). Although they have some chemical defences (Ostrofsky and Zettler, 1986), which may be induced in response to grazing damage (Jeffries, 1990b), these defences have not saved terrestrial plants from attack. It may simply be one of these scale problems again (Fig. 6.10). The major herbivore groups, especially of insects, have not moved into freshwaters on the evolutionary time scale. Major groups of freshwater invertebrates, including insects, were apparently well established before most modern macrophytes invaded the habitat. (Wootton, 1988). The herbivores, already using detritus and microflora, may not have adapted to exploit this novel food source.

Table 6.2 Alkaloid content of leaves of freshwater macrophytes, given as mg alkaloid per gram dry weight of leaf. Data are means ± one standard error. Alkaloid compounds are credited with being anti-herbivore defences, so these macrophytes appear to possess chemical defences, even though the levels are low compared to terrestrial plants. (After Ostrofksy and Zettler, 1987)

Plant species	Alkaloid content
Ceratophyllum demersum	0.24 ± 0.08
Elodea canadensis	0.19 ± 0.04
Myriophyllum spicatum	0.28 ± 0.07
Potamogeton crispus	0.56 ± 0.28
Potamogeton natans	0.27 ± 0.17

6.5 Freshwater communities and trophic structure

Throughout the sections on competition, predation and herbivory there are recurrent general patterns in the ways animals feed. Some are predators, some graze the algal films off rocks or plants, others chew up larger detritus. These categories are not simply trivial terms. The patterns and distributions of feeding mechanisms are a revealing aspect of freshwater ecology and allow us to explore communities in a way that looks at the functional trophic groups to which animals belong, in addition to taxonomic classifications.

The basic mechanisms of feeding have been drawn up into five categories (Cummins 1973, 1974; Lamberti and Moore, 1984).

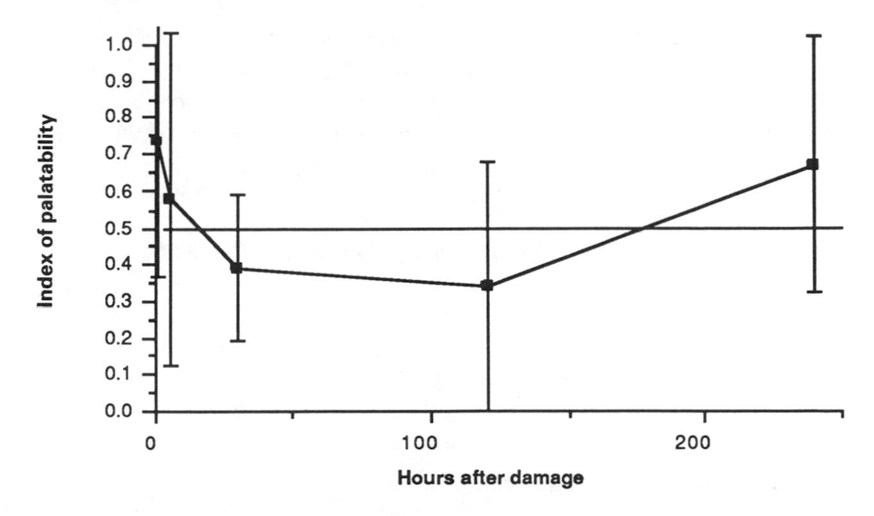

Fig. 6.10 Evidence of induced defences in a pondweed. The graph shows the change in palatability of damaged leaves of a pondweed, *Potamogeton coloratus,* compared to undamaged leaves, at intervals following damage to some of the leaves on individual plants. Larvae of the caddis *Trianodes bicolor* were used as the grazer. The index of palatability is the amount of damaged leaf eaten divided by the amount of damaged leaf plus undamaged leaf eaten, for pairs of damaged and undamaged leaves. If leaves were equally palatable the index will measure around 0.5. There is a decline in the apparent palatability of damaged leaves suggesting some change perhaps in chemical defences, induced by the damage. Data points are mean ± standard deviation. (After Jeffries, 1990)

1. *Shredders* These chew, mine, bore and gouge large particles such as leaves, stems, branches which may be alive or dead.
2. *Grazer/scrapers* These graze and scrape the periphyton off other surfaces.
3. *Collector/filterers* These filter particulate matter, alive or dead, from the water.
4. *Collector/gatherers* Again feeding on small particulate matter but gathering the fine detritus off the sediment or other surfaces rather than from the open water.
5. *Predators* These are subdivided into engulfers, which eat the whole prey item swallowing it whole or chewing, and piercers, which pierce the prey and suck fluids out.

Besides the general descriptive use of such terms they are another way to look at the patterns in communities. Food sources vary along the length of a river or with the age of a lake. Narrow upland streams rely heavily on allochthonous leaf debris which can be used by shredders. Downstream the fine debris of demolished leaves will support collectors. Changes in functional groups reflect changes in food sources, nutrient processing and energy flow. This has important practical implications. Human activity may dramatically alter food sources, whether it be leaves off trees or particulate debris in effluent. How we alter the type and amount of food input will affect the trophic groups.

6.6 Movement, migration and colonization

Most animals move around, even sedentary creatures, when the substrate they are on shifts or at stages in their life cycles adapted for dispersal. The comings and goings add further complexity to the community patterns in space and time. In freshwaters three different movement phenomena play important roles:

1. Colonization and recolonization of habitats.
2. Drift, a phenomenon of lotic systems when normally benthic creatures intentionally let themselves be carried along in the current.
3. Migrations, whether the huge journeys made by fish such as the salmon, *Salmo salar* L. and the common eel, *Anguilla anguilla*, back and forth to their breeding grounds or the movement of lentic animals up and down through the water column following a daily cycle.

Plate 20 The March Brown, *Rhithrogena haarupi*, a mayfly of moderately swift-flowing rivers. This stage in the fly's life is referred to as a 'dun' just after it has emerged from the nymphal stage. It will shed its skin again to become a 'spinner' after which it is ready to reproduce and lays its eggs over water. (Photo: Derek Mills)

Colonization can be studied at many scales, from the invasion of continents to arrival at individual water bodies and across time scales of geological epochs to minutes. Community ecology has concentrated on the year-to-year dispersal of species and how this interacts with other processes to shape communities. Many freshwater insects have flying adult stages and disperse very well. Dragonflies can disperse across continents and bugs and beetles are strong flyers. Flights may be synchronized to cues such as temperature and end in response to humidity and light reflection by the water's surface. New habitats are rapidly colonized by these taxa. The arrival of species over time depends on the habitat's size, proximity of sources of colonists, and survival of the arrivals as predation, competition, facilitation and microhabitat needs influence the community. The theoretical background to such processes has been extensively developed as island biogeography theory, (MacArthur and Wilson, 1967). Even very slow colonists will eventually arrive. For example, flying adult beetles have been caught with pea mussels (*Pisidium* spp.) caught around their legs (Kew, 1893; see also Fig. 6.11). *Pisidium* are not the most active animals when under water so their hitch-hiking may be an important, if risky, way of dispersal. Alternatively, some members of generally mobile groups have become flightless. Permanent habitats may encourage this. Many small ponds and streams are not at all permanent on the time scale required to fix flightlessness as a character, but animals of small habitats in areas where dispersal may be extremely risky may still become flightless, even if they retain wings. This seems to apply to surface skimming bugs on some upland streams. Sheldon (1984) reviews colonization dynamics of aquatic insects. Minshall *et al.*, (1983a) describe recolonization of streams using island biogeography theory.

Adult insects of lotic waters commonly show a strong upstream flight pattern, rather than dispersing widely. By doing so they are compensating for the downstream drift that is found in many of their larval stages.

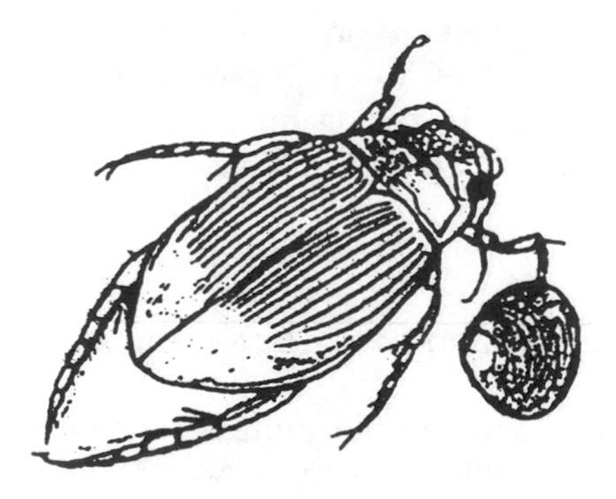

Fig 6.11 Adult great diving beetle, *Dytiscus* spp., with freshwater bivalve, *Pisidium* spp., holding on to leg. (After Kew, 1893)

Generally, stream benthos are adapted to resist the current and are not swept up into the open water where they are vulnerable to fish. However, drift occurs under certain conditions. Typically there is a diurnal rhythm with drift occurring in the early hours of darkness. This may represent a way of spreading and colonizing the stream microhabitats, especially for the newly hatched larvae. Invertebrates will also move away from disturbance be it natural, such as a predator, or artificial, such as pollution. Upstream movement also takes place over the substrate and by burrowing through it. Complex integration of recolonization and life history patterns are associated with temporary habitats such as ephemeral ponds and are described by Williams (1987).

Migrations may also be linked to life history, notably the famous journeys of salmonids and the eel. Many salmonids breed in freshwater, migrate out to sea and return after a period of feeding and growth, a pattern called anadromy. Conversely, the eel breeds out at sea, in the Sargasso in the western Atlantic. The precise routes and circumstances are still a mystery but the young elvers migrate to freshwater to grow to maturity, this arrangement being catadromous. Presumably such migrations have developed to benefit the adult and juvenile stages, though the very long journeys involved these days may be an artefact of continental drift pulling the Atlantic Ocean ever wider. Vertical migrations are commonly found in lentic plankton and pelagic animals (Zaret and Suffern, 1976; Stich and Lampert, 1981). Typically, species ascend at night, descend at dawn. There may be many causes, such as ectotherms adjusting to changing thermal regimes, or zooplankton grazers responding to algal migrations that reflect shifting nutrient concentrations, but the most compelling cause appears to be predation. The exposed, pelagic waters are very hostile, especially due to visually hunting fish. By descending during the hours of light and ascending at night to graze the algae zooplankton are reducing the risk. Such rhythms can develop in just a few years in zooplankton populations of previously fishless lakes, lacking migration, once planktonic fish have been introduced. The few individual zooplankton organisms that migrated would be very rapidly selected for the intense fish predation. Previously, migrating individuals would be at a disadvantage compared to non-migrating compatriots that stayed up in the productive euphotic zone all the time (Fig. 6.12).

6.7 Disturbance, chance and stochastic events

The preceding sections may give the impression that communities are neat, predictable assemblages in a deterministic world. If only we could quantify all the constituent parts and processes then we could predict how each community will form and progress. Thankfully life cannot be reduced to this cheerless predictability and all communities have a degree of chance to their

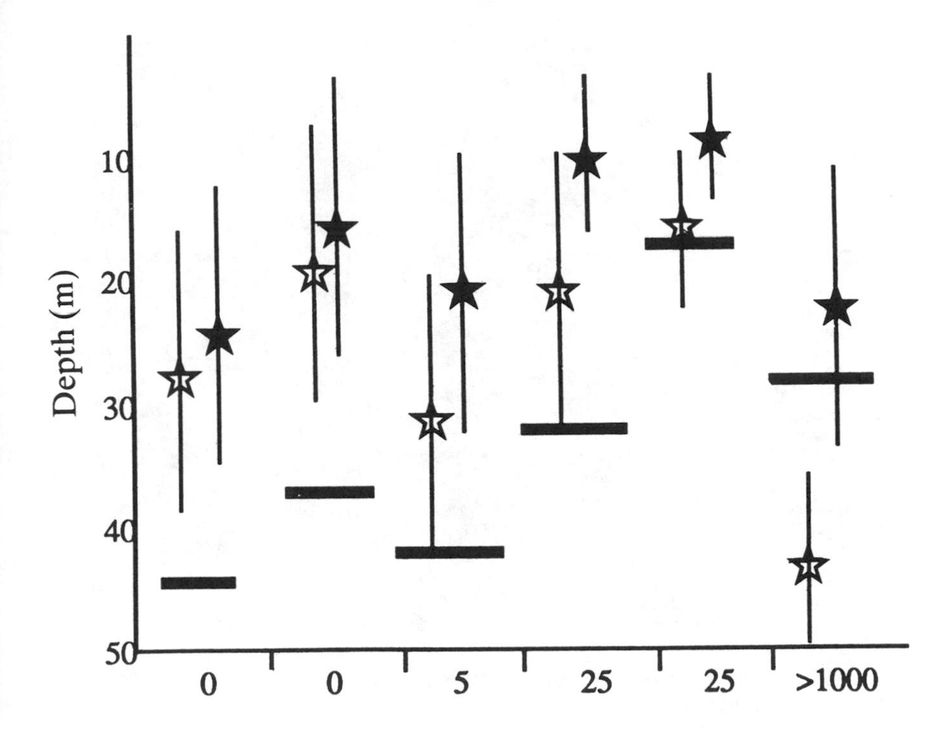

Fig. 6.12 Zooplankton migration as an anti-predator defence. Depths of *Cyclops* zooplankters, in six Polish lakes exposed to fish predation for different periods, during daylight and at night. The numbers of years since fish were first introduced to the lakes are given along with mean depth (± 1 standard deviation) of *Cyclops* during daylight ☆ and darkness ★. Thick horizontal lines mark the level at which light is 1 per cent of surface incident illumination. These data are for August. Diurnal migrations are greater the longer populations have been exposed to fish predation. (After Gliwicz, 1986).

structure and dynamics.

The predictability of an event depends on the scale involved. Some events may be very regular but on a huge time scale. The impact of extra-terrestrial objects on Earth is a current favourite with an apparent regularity of some 26 million years. On the other hand most freshwater habitats do not exist for this long. If such an event occurs during their lifetime, predictable though this is in the long term, there is no sense of rhythm or regularity to the event as a process in freshwater systems. It is a one-off event. Some events, big or small, do not fit into basic patterns of probability and are called stochastic. Chance must play some part in wildlife populations too. No animal or plant can determinedly ensure that it colonizes all suitable habitats in an area. Sometimes arrivals or absences might be sheer chance (Talling, 1951; Jeffries, 1989) but these unpredictable events might have important

Plate 21 The brown trout, *Salmo trutta*, a fish inhabiting a wide range of lakes and rivers holding water of good quality (i.e. high dissolved oxygen and low water temperature). The fish depicted here is from a mountain stream in the Austrian alps. (Photo: Derek Mills)

consequences through the combinations of plants and herbivores, predators, prey and competitors that are assembled.

Disturbance, such as floods and spates, droughts, natural pollution, whether stochastic or regular, will also keep upsetting the deterministic progression of a community, for example disturbance as a factor in streams (Stanford and Ward, 1983). The result is that communities may never reach a stable form. They may always be in a form that lacks any equilibrium.

Community ecology has recently gone through a period of intense reappraisal as deterministic and non-equilibrium processes have been brought together. These topics remain comparatively mysterious and ripe for research. Hildrew and Townsend (1987) provide a very good integration of deterministic and non-equilibrium ideas in benthic communities. Several excellent general ecology textbooks now exist which describe the curernt views on abiotic and biotic, deterministic and non-equilibrium community ecology (Strong *et al.*, 1984; Begon *et al.,* 1986; Diamond and Case, 1986; Gee and Giller, 1987).

7 Eutrophication

Pollution of freshwaters is such a familiar idea that we may lose track of the bewildering variety of causes and consequences. This chapter and Chapters 8 and 9, explore many of the problems and some of the answers. To define what a pollutant is in precise physical or chemical terms is impossible, and naive. Some pollutants such as oil or poisonous chemicals readily spring to mind, but others are less obvious such as hot water from industry, or under some circumstances, very important parts of the natural chemical cycles, such as plant nutrients, are not at all what the term 'pollution' conjures up as human disturbance. A workable definition has to embrace a kaleidoscope of pollutants and many levels of damage. For example, to simplify Holdgate's (1979) definition: 'The introduction by humans into the environment of substances or energy that damages life or the physical environment'.

Anything, in the wrong place, in damaging quantities, can be a pollutant. Freshwater pollution can usefully be divided into two broad categories and appreciating the difference can help us to understand, prevent and reverse the damage. Firstly, there is pollution that is caused by an imbalance, a perturbation, of important natural processes. There are two planet-wide forms. One is an imbalance in nutrient processes, in particular a destructive excess, called eutrophication. Secondly, a perturbation of the natural acidification processes of the hydrological cycle, resulting in the phenomena of acid rain, black snow and freshwater acidification. Eutrophication and acidification are the subject of this and the following chapter respectively.

Other pollutants represent entirely unnatural, novel impositions on the ecosystem. For example an oil spillage into a river, an overspraying of insecticide into a lake, a flux of heavy metal ions from mine workings are all abnormal additions. They do not have a natural place in ecosystems. The diversity of these pollutants is described in Chapter 9. Nature can be destructive, such as the deluge of volcanic debris into the rivers and lakes around Mount St Helens in North America in 1980 (Baross *et al.*, 1982) or the extraordinary conditions of heat and chemical stress in thermal springs, but these are a natural part of the planet's life. Perhaps the only factor that really unites the variety of pollutants is that they are all the result of human activity: they are anthropogenic.

7.1 Defining eutrophication

In Chapter 1 the nutrient status of lakes, especially the role of nitrogen and phosphorus, was described. There is a perfectly natural span from nutrient-poor (oligotrophic) to nutrient-rich (eutrophic) waters, both lentic and lotic. The term eutrophication can refer to the natural enrichment of a lake over time, or river along its course, but the word has increasingly assumed a definition that primarily refers to a form of destructive pollution. Eutrophic waters are nowadays commonly thought of as those showing signs of excess nutrient loading with associated changes in flora and fauna. It is impossible to put neat thresholds or measures to describe what is or is not eutrophication. A polluted, eutrophic water in an area of naturally nutrient-poor waters may have a smaller nutrient load than an undegraded water in a naturally nutrient-rich catchment. Eutrophication is a form of damage that must be gauged relative to the natural balance that should prevail in the area and the size of the basin. Vollenweider (1968) suggested that definitions of lake trophic status should rely on a combination of depth and phosphorus loading (see Fig. 7.1). Nor do eutrophic waters necessarily have high levels of nutrients present all the time. The nutrients may be gleaned and utilized by the plants, especially algae, so quickly that amounts at any instant are undetectable. It is the rate at which they are available, the flux of the loadings that is important (Table 7.1).

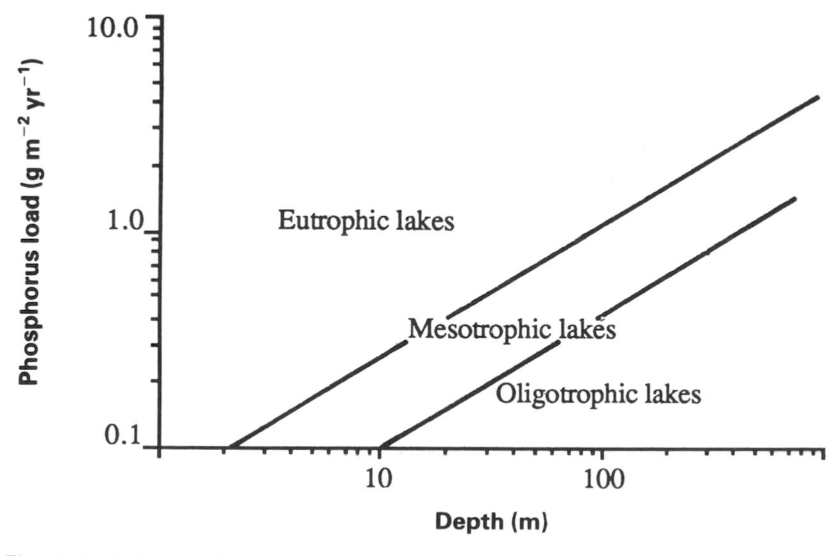

Fig. 7.1 Lake trophic status as a function of depth and phosphorus loading. (After Jorgensen, 1980)

Table 7.1 Trophic classification of lakes. (After Jorgensen, 1980)

	Characteristic				
Trophic status of lake	Mean primary productivity	Phytoplankton biomass (mg C m^{-3})	Total organic, carbon (Mg l^{-1})	Total P (μg l^{-1})	Total N (μg l^{-1})
Dystrophic	<50–500	<50–200	3.0–30	<1.0–10	<1.0–500
Oligotrophic	50–300	20–100	<1.0–3.0	<1.0–5.0	<1.0–250
Mesotrophic	250–1000	100–300	<1.0–5.0	10–30	500–1100
Eutrophic	>1000	>300	5.0–30		
Hypereutrophic				30–>5000	500–>15 000

What characterizes eutrophication is not just the supply of nutrients but also a set of linked biological consequences. Eutrophication as a pollution problem is a mixture of both the causes and symptoms. The simplistic oligo- to eutrophic classification was devised by Weber (1907) for bog soil solutions, then developed by Naumann (1919) to describe lakes. Since water chemistry analyses were inadequate at that time, the scheme was designed to reflect the obvious differences in phytoplankton, which in turn reflected trophic status, between lakes. So the early classification embraced cause and effect. Subsequent attempts to define precise thresholds have proved difficult, it is the dynamics of the whole system that are important. Eutrophication is a problem of the rate at which change happens. It is an acceleration of natural ageing and enrichment, though the degraded communities that result probably represent genuinely different communities and not the natural progression that would have arisen at normal enrichment rates as a lake ages. The community has been switched to a different form. A trophic equilibrium exists as long as environmental conditions stay the same or change at a natural rate, the water maturing into diverse, productive old age. Jorgensen (1980) suggests that eutrophication occurs after the habitat's ability to buffer against excessive nutrient inputs, especially the phosphorus binding capacity of the sediments, is exhausted. The degradation can then happen quickly and represents a definite shift in the state of the ecosystem, not merely an accelerated gradation. These problems have prompted some scientists to wish that the naive oligotrophic–eutrophic idea could be abandoned (Moss, 1980b). Fertilization and enrichment have been suggested as alternative terms for the pollution form of eutrophication but have not caught on. Cultural eutrophication is also used and has the useful implication that this is an anthropogenic process.

Since neat definition are difficult and dangerous it is better to describe the typical pattern, the causes and symptoms of eutrophication.

7.2 Eutrophic ecosystems

The input of excessive plant nutrients starts a chain of events which follow a typical pattern. Not all happen in all eutrophic waters, but the general pattern is common:

1. Changes in algae, both phytoplankton and periphyton, occur. Species composition and productivity alter, the latter often greatly elevated. Blooms of species able to exploit the conditions break out. The water becomes turbid, supersaturated with oxygen in the daytime, perhaps anoxic at night and during decay of blooms when populations crash. Some species produce toxins, notably blue–green algae which become increasingly dominant as eutrophication proceeds.
2. Nutrient input alters the macrophyte communities. Initially, some

tolerant species able to use the increased nutrients may flourish, assuming nuisance proportion. Sensitive species are lost and diversity declines. Increasing turbidity, algal blooms, anoxia and sediment changes eventually harm even tolerant species. Severe macrophyte loss with only a few emergent species, or none at all, characterizes badly eutrophic habitats.

3. Sediment changes. Nutrient input may happen in conjunction with increased solid effluent that accelerates sedimentation. Death of phytoplankton blooms, anoxia and macrophyte decay all add to this and an unstable, anoxic, semi-liquid mud is a common feature.
4. The animal life changes with a general loss of diversity, though a few species may flourish in the enriched conditions and their productivity can be high. The losses will be partly due to changes in water chemistry and anoxia, but also due to the alterations in the algae and macrophytes. Loss of physical cover provided by the latter is a major factor and once a few species start to decline foodweb systems can break down catastrophically. Oligochaete worms and chironomid midge larvae dominate the remaining fauna. Fish species are progressively affected, with loss of sensitive game fish and only a few, if any, coarse fish left. In extreme cases (hypereutrophic waters) regular fish kills due to erratic oxygen and pH regimes are a characteristic feature.
5. Besides the biological changes the entire look, or amenity value, conspicuously alters. Badly eutrophic waters acquire a thick, pea-soup look. Noxious muddy odours develop from cyanophytes and actinomycete fungi, which can affect the surrounding environs. Severe health problems can arise. Some algal toxins affect humans who swallow water while swimming. A bacterium, *Clostridium botulinum,* flourishes in the anoxic sediments. This is a widespread organism but when populations bloom toxins which affect birds and mammals build up, the typical paralysis that results is known as botulism. Excessive nitrate levels in drinking water are a health risk. Adults are vulnerable to carcinogens called nitrosamines, which form from amine and nitrate compounds. Large inputs of effluents increase the quantities of these precursors. Babies are vulnerable to a condition called methaemoglobinaemia ('blue baby syndrome') in which nitrogen compounds absorbed from drinking water bind to blood haemoglobin in place of oxygen, with a risk of suffocation.

It is worth pointing out that in some instances eutrophication can be used constructively. Intensive fish-farming in many tropical countries relies on tolerant species farmed in intensively fertilized waters. The massive productivity is turned to advantage.

7.3 Eutrophication: an example

Eutrophication has been studied all around the world, yielding many and varied examples, though all reflect the common pattern. Some examples have almost become clichés, but are none the less well worth their use, having been so thoroughly researched and illustrating so many points. The Norfolk Broads, in eastern England is just such a classical example with the causes, consequences and some recent attempts at cures all illustrated (Mason and Bryant, 1975; Moss 1977, 1980a, 1983; Moss *et al.*, 1986).

The Broads are a unique system of shallow lakes and associated rivers that ramify through north-east Norfolk. They often seem to form a network of waterways and lakes, called Broads, having a parallel existence to the surrounding roads and villages but mysteriously separate. What are now the Broads were dug out from dry land for peat for 300–400 years prior to abandonment and subsequent flooding in the thirteenth century. Climatic change, land subsidence and sea level rises all contributed to this process though the fens were still used as a source of reeds, used in building and thatching, and as a transport system. From the late eighteenth century onwards pressures started to encroach, slowly at first but accelerating towards the end of the nineteenth century. The local population increased, with resulting increase in sewage output especially from the large city of Norwich. Intensive use of the rich agricultural land all around increased and areas of adjacent fen were drained. Added to all this, a thriving holiday industry grew up, based on pleasure boats, from the late nineteenth century, again releasing sewage, eroding banks and stirring sediment. The establishment of feral coypu, escapees from fur farms, may have exacerbated the problem of bank damage. These developments caused an increase in nutrient loadings and turbidity and changes in the habitats started with severe eutrophication damage evident from the mid-twentieth century. Only recently has there been any improvement and this is largely restricted to a few sites where direct intervention has been attempted. The changes have been traced partly from the natural history literature, especially macrophyte loss. Palaeolimnology has provided good evidence (Figs. 7.2 and 7.3). This is the study of the past history of freshwaters, commonly using cores taken from the sediment. Sediment structure, chemistry and remains of animals, such as hard insect exoskeleton, and plants, particularly algal cell walls, can be analysed. Past nutrient budgets can sometimes be tentatively devised. Palaeolimnology using sediment from the Broads suggests they were originally calcareous, marl-rich waters, with little phytoplankton though abundant epiphytic algae, suggesting a necessary abundance of macrophytes to live on, of which the macrophyte alga *Chara* was common. Sedimentation rates were low, less than 1.0 mm year, and phosphate levels around 20 μg total p–L^{-1}, were moderately rich, in keeping with the natural productivity of the area (Fig. 7.4). Anecdotal nineteenth-century works suggest wildlife flourished with exploitation of

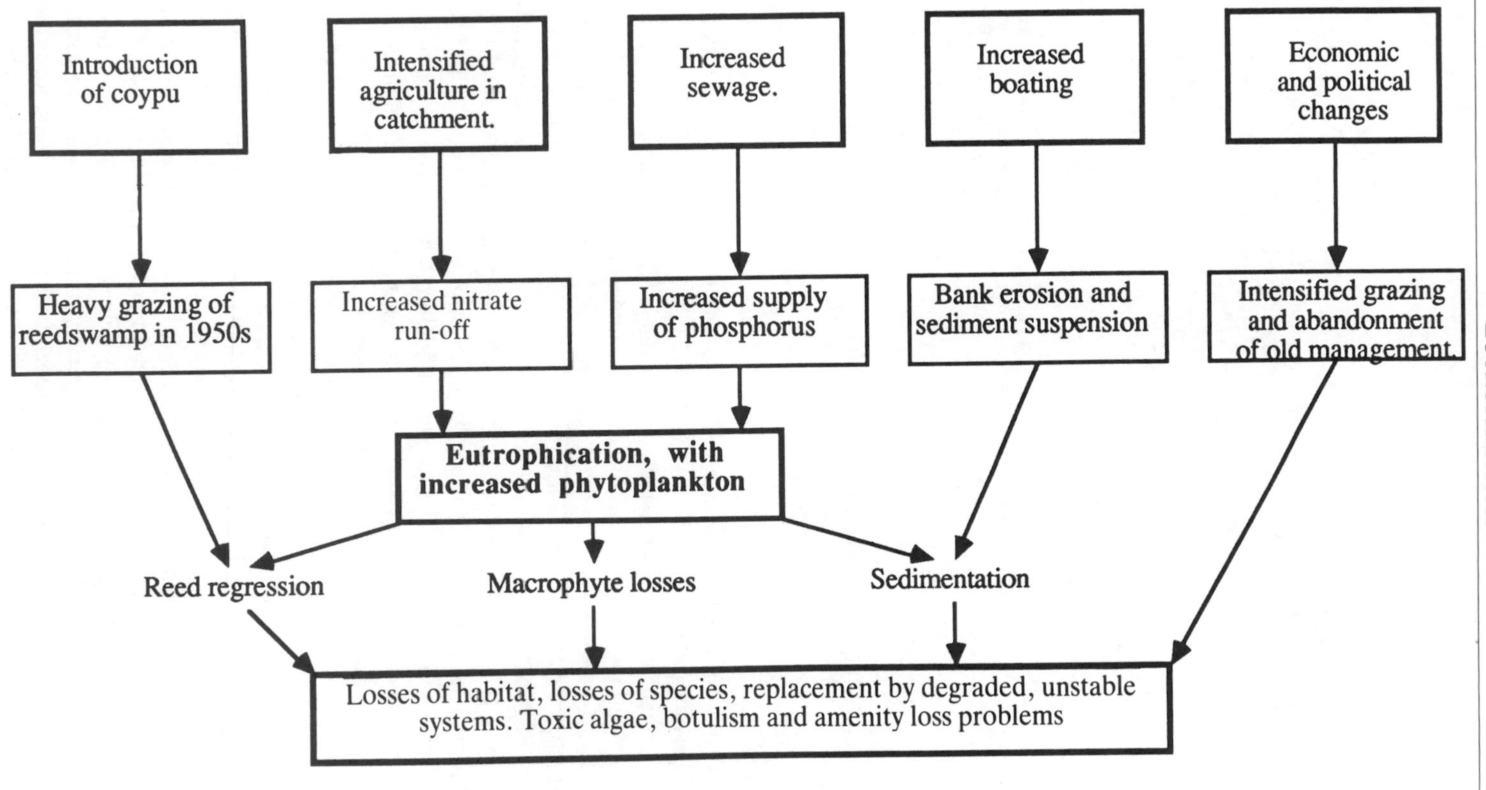

Fig. 7.2 Main cause and effect relationships for eutrophication of Norfolk Broads. (After Moss, 1983).

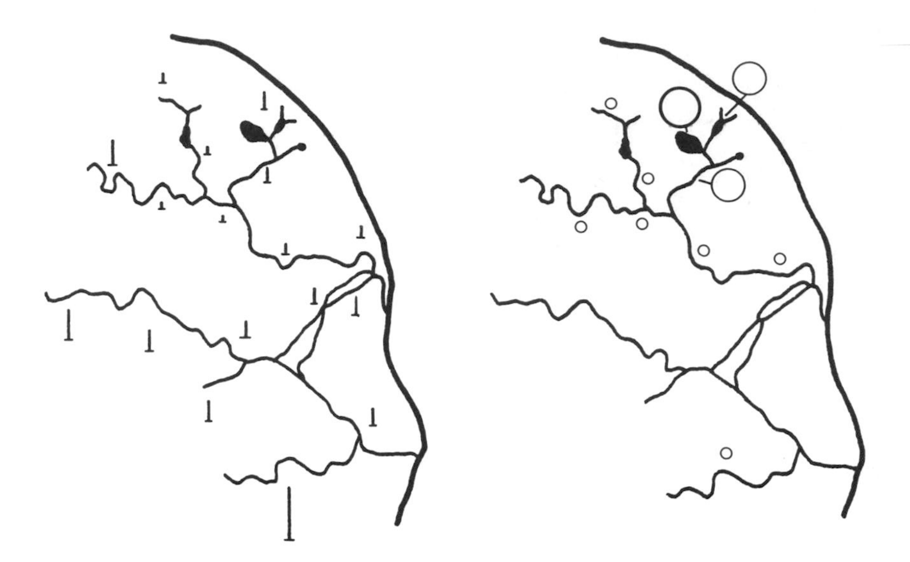

Fig. 7.3 Summer Secchi disc transparencies and colonial blue–green algae populations in the Norfolk Broads. (After Moss, 1977)
algae ◯ < 40,000 ○ 5000–20,000 ∘ < 500 Secchi depth | = lm

fisheries and wildfowl but with a hint of enrichment in the dense weed growth in some rivers.

During this century changes have been drastic. It is worth remembering that such shallow waters would fill in and change naturally but the rate of change and community degradation represent a qualitative break with the natural sequence. Macrophyte loss has been a most visible change. Reed swamp has died back since mid-century. Estimates from some Broads suggest only 40 per cent is left and reed quality is said to have declined, as extensive nutrients have allowed rapid but frail growth. Water lily fringes have been lost and patches only grow sporadically. Submerged macrophytes have been decimated. Two main communities have been recognized, a charaphyte sward, perhaps the initial stage and a taller mix of milfoil (*Myriophyllum*), hornwort (*Ceratophyllum*), with lilies, emergents and associated algae (Mason and Bryant, 1975). Typically up to ten macrophytes were found in the 1920s–1930s and these remained until the 1940s with some evidence of choking or initial losses of the most sensitive species. Surveys since the 1960s all document massive loss. Records for the Broads off the Rivers Ant and Alderfen in 1968 suggest only floating leaved species were left. In 1972–73 a survey of 28 broads revealed 10 devoid of macrophytes, 12 with very few, generally lilies with floating leaves, and 6 with reasonably intact communities. In the 1980s another survey, this time of 42, gave 4 with the *Chara* community, 6 with the milfoil/hornwort type and algal blooms in

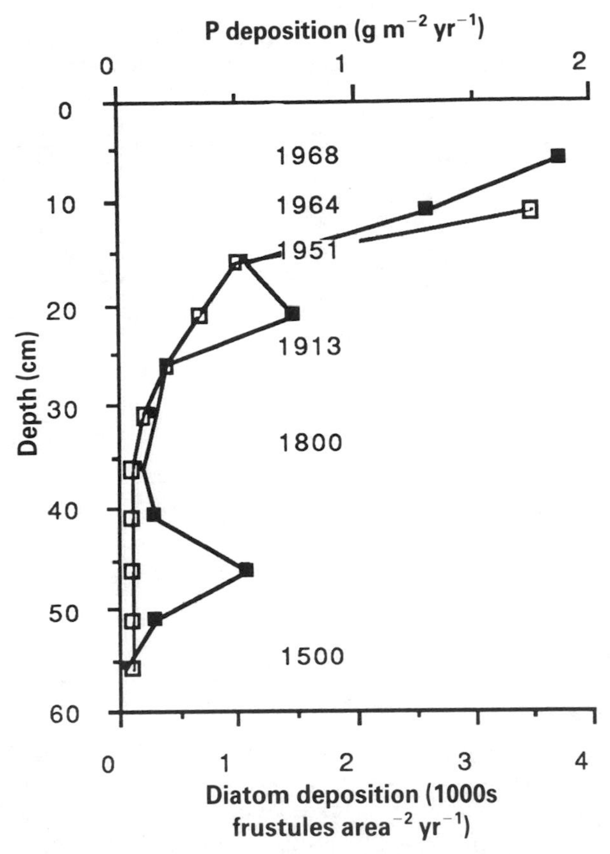

Fig 7.4 Deposition rates of phosphorus and diatoms in sediments of Alderfen Broad, based on studies of sediment cores. Approximate dates are shown. Note the increase in phosphorus from the early twentieth century onwards and increasing diatom numbers. The twentieth-century diatoms are largely phytoplankton species, the earlier peak (around 1650) is entirely epiphytic species perhaps recording some natural change in the ecosystem. (After Moss *et al.*, 1979) ■ = Diatoms □ = phosphorus

the remaining 32. Not only had macrophytes changed, phytoplankton populations were high, with certain taxa such as blue–green algae dominant (Table 7.2). Zooplankton, once diverse and dominated by taxa associated with weed beds such as the waterfleas *Eurycercus* and *Simocephalus*, had changed to open water types such as *Daphnia, Bosmina* and the copepod *Diaptomus* with small rotifers dominant. Loss of macrophytes may have exposed the zooplankton to increased fish predation causing this shift to smaller species. Benthos has altered from a diverse fauna to domination by worms and midges, again loss of macrophytes is probably a major factor. Very specific damage may be done. The reproduction of a rare dragonfly, the Norfolk aeshna, *Aeshna isosceles* (Muller) is closely tied to the presence of the water soldier plant, *Stratiotes aloides,* into which the female dragonfly

Table 7.2 Macrophytes of selected Broads, surveyed in 1972–73. The abundance of macrophytes present is given by the scale 1 = very scarce, 2 = scattered patches, 3 = abundant, 4 = dominant. (After Mason and Bryant, 1975)

Broad \ Plants	*Chara* spp.	*Fontinalis antipyrectica*	*Nymphaea alba*	*Nuphar lutea*	*Myriophyllum spicatum*	*Ceratophyllum demersum*	*Hippuris vulgaris*	*Callitriche stagnalis*	*Potamogeton pectinatus*	*Zanichellia palustris*	*Najas marina*
Cockshoot						None					
Surlingham						None					
Barton			1	1							
Alderfen			1	1		1					
Horsey mere	2				3				3		
Hickling	2	1			2	1			4	1	1
Upton	1		2	1					1	1	4
Calthorpe			3	3		1	3	4			

implants her eggs. The plant is sensitive to eutrophication and loss of the plant endangers the dragonfly. Fisheries have remained productive, but shifts in age structure of populations occurred with fewer older fish and rapid growth of young due to planktonic productivity. Fishkills happen, associated with the alga *Prymnesium parvum* and botulism poisoning regularly affects birds. Some bird species may have declined, such as the bittern as a result of the reed die-back, or the coot, which feeds on plants and invertebrates, from macrophyte loss. Overall the whole impression is of degradation and loss with destructive imbalances in the physico-chemical environment and surviving populations.

The changes in wildlife are due to underlying physical and chemical alterations. The crucial change has been an increase in phosphorus and nitrogen loads in the Broads system. Calculations of original levels suggested 10–20 μg total Pl^{-1} and in some undegraded Broads, for example Upton, levels are similar today, 18–39 μg Pl^{-1} and 0.0–1.7 mg NO_3^{-1} (Moss, 1983). Measures like this can be deceptive as eutrophic waters may have little free phosphate due to the rapid uptake but, for comparison, eutrophic Broads show ranges of 50–1000 μg P and 0.0–3.2 mg NO_3^{-1}. Moss (1986a) provides estimates of concentrations and total loadings of phosphorus for Barton Broad, based on palaeolimnological work, over two centuries: in 1800 a concentration of 13.3 μg Pl^{-1} and phosphorus loading of 0.4 g m^{-2} yr_{-1}; in 1940 concentration of 119.0 μg Pl^{-1} and phosphorus loading of 3.6 g m^{-2} yr^{-1}. Measured either way this is a nearly ten-fold increase.

Associated problems have been saline water tables and acidic flushes as drainage has exposed peat deposits. Sediments have become unstable, with a seasonal stirring up by boats in the holiday areas and anoxic mud.

Recent experimental evidence suggests the interactions of nutrients, phytoplankton, macrophytes, zooplankton and planktivorous fish may be very complex. (Balls *et al.*, 1989; Irvine *et al.*, 1989; Stansfield *et al.*, 1989). Either macrophyte or phytoplankton dominated systems can occur across a wide range of nutrient loadings. The switch from one community to the other can be brought about by various mechanisms. In particular the survival of effective zooplankton grazer populations, preventing algal blooms, may be important. Pesticide damage to zooplankton communities may have caused degradation (Stansfield *et al.*, 1989).

This mass of symptoms can become bewildering after a while but they actually make up an integrated set of causes and symptoms. Philips *et al.*, (1978) have devised a basic mechanism for the degradation, especially macrophyte loss, which not only shows how the progressive changes arise but also the links between different parts of the communities (Fig. 7.5). Note how both alternative states, the undegraded and the eutrophic, are self-reinforcing. Once one or the other starts, then positive feedback promotes that state. Also note how one event can precipitate several others, for example macrophyte loss freeing algae from competition, exposing invertebrates to fish predation and anoxia by decay.

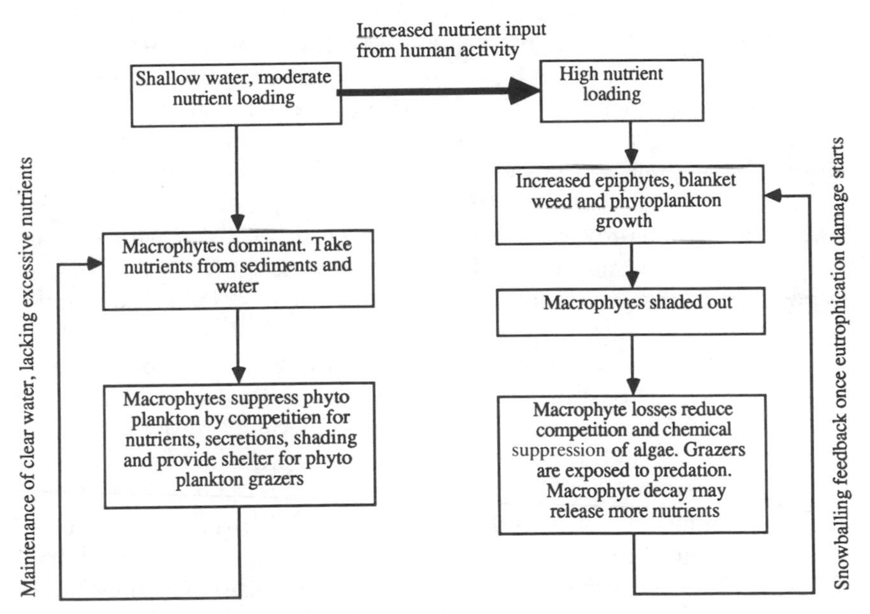

Fig. 7.5 Mechanism to account for the loss of macrophytes due to eutrophication. Note how unpolluted and eutrophic ecosystems are self-sustaining. The pollution is perturbation that switches the ecosystem across to a new state. (after Philips *et al.*, 1978)

The Broads are a complex and fascinating example, but there is the same basic pattern in eutrophic waters all over the world.

Nutrient load increases - - - - - → Macrophyte loss- - - - → Algal blooms,
and algal changes invertebrate decline, anoxia and toxins

7.4 Sources of excess nutrients

For all their geographical and biological variety the general pattern of eutrophication is the same in many water bodies. It is the unnatural elevation of phosphorus loading that is the main cause of the switch to the self-sustaining process of degradation. The diversity of phosphorus and nitrogen sources around the world explains the global extent of the problem and the difficulties of containment and cure. A simple division of sources into diffuse versus point (or discrete) inputs is a useful step as recognition of the type of input is important for prevention and cure. A point source refers to nutrients released at one discrete input. These are often very obvious such as the drains from a milking parlour or a sewage effluent outfall pipe. Diffuse sources are those where the input occurs widely over an area such as wind-blown fertilizer deposition or a nutrient-enriched water table.

1. *Sewage* A major source of nutrients worldwide, whether it be massive urban discharges or local point sources such as poorly maintained septic tanks. Sewage treatment plants originally were designed to alleviate major pathological dangers such as disease organisms and remove gross organic pollution. These are called primary and secondary treatments. The effluent released may be safe but is still laden with nutrients, perhaps in a more accessible form after the breakdown. Tertiary treatment has been developed now which is designed to strip out, or 'polish', the phosphorus nutrients before effluent is discharged, but it is not regularly installed.
2. *Fertilizers* Agricultural fertilizers, largely nitrogen and phosphorus, are designed to boost plant productivity. Overspraying into water and run-off following rain simply dump the nutrients into freshwater rather than on to crops. Nowadays a more insidious problem has developed. Fertilizer nutrients have started to reach the soil water table and deep aquifers. This has taken years but equally, as nutrients slowly leach down the input to aquifers will continue for many years even if we stopped spraying at the surface, which we have not. This source is very difficult to tackle. Waterways in intensively farmed areas have started to degrade from this diffuse source and drinking water supplies from this once clean source are vulnerable to excessive nitrogen levels. Wind-borne fertilizers are a problem in some countries. Particular attention

has been paid to this source in the Netherlands where inputs may be causing eutrophication at some sites and acidification at others, through the nitrogen cycle (see Chapter 9).

3. *Detergents* These used to be rich in phosphorus compounds, part of the chemical agents that prevented dirt settling back on to the clothes once they had been washed. Domestic detergent waste has been credited with causing eutrophication in lakes in several developed countries. New 'ecologically sound' detergents are designed largely to avoid this. Industrial detergents are another source.
4. *General urban run-off* The run-off from roads and built-up areas is a fearsome mixture of pollutants but includes a large proportion of nutrient-rich, organic debris. Waterways through urban zones and ponds and lakes therein are commonly eutrophicated.
5. *Point sources* A vast variety of unlikely point sources can cause severe local damage. Intensive agriculture facilities are a common problem, for example milk parlours, battery animal houses, abattoir run-off and silage leakage. In many cases, notably silage, the degree of enrichment is so severe that gross-organic pollution develops which goes beyond even the eutrophic state.
6. *Flooding* The building of reservoirs by flooding of the countryside results in a massive flush of nutrients, from soil and decaying plants, into an immature ecosystem and often with anoxia developing in the deep water as the terrestrial debris rots. This may cause reservoirs to be eutrophic from the moment of their creation.
7. *Wild animal faeces* The droppings of birds are rich in phosphorus, so much so that the colossal deposits of seabird guano off South America sustained a profitable quarrying industry for fertilizer. Any concentration, especially nesting or roosting, of birds on freshwaters can result in heavy inputs, The Norfolk Broads, inevitably, provide an example. Moss and Leah (1982) suggest that a winter roost of black-headed gulls accounted for 53–72 per cent of maximum total phosphorus concentration in Hickling Broad. Systems enriched in this way have been charmingly entitled guanotrophic.

7.5 The restoration of eutrophic waters

The abundance and diversity of eutrophic waters has resulted in many approaches to prevention and cure. A set of general principles can be discerned.

1. *Cut the supply of nutrients into a system*
   (a) This can be achieved by cutting down on diffuse sources, for example use of fertilizers, treatment of a point source before discharge or treatment of water in a system that is already heavily

loaded by abstraction, stripping of nutrients and then releasing the water back into the system.

(b) Isolation of water bodies from nutrient-rich inflows can be tried in some cases. This applies especially to lakes where nutrient-laden inflows cannot be treated.

(c) Provision of an entirely new water source, for example from a new borehole.

2. *Remove or reduce nutrients already in the system*

   (a) This usually means physical removal of nutrient-rich sediments that act as a source of resupply and maintain the self-reinforcing cycle of degradation. It might also involve removal of macrophytes, and occasionally other life, using the wildlife as mechanisms for extracting and binding up of nutrients which can then be taken out and away (Livermore and Wunderlich, 1969).

   (b) Nutrient levels can also be reduced by disruption of the resupply, through dilution with clean water, aeration to prevent anoxia and management of stratification. Essentially all these interventions are designed to restore abiotic conditions that alleviate eutrophication.

3. *Restoration of biotic conditions that promote a non-eutrophic state* If eutrophication represents a switch to an alternative state from the natural progression it may be possible to switch conditions back by manipulating the animals and plants. At the heart of this is an attempt to encourage the non-eutrophic macrophyte–algal balance. Such techniques are called biomanipulation.

The theory ought to apply to both lotic and lentic systems, but in practice rivers, flowing through endlessly changing catchments with many input sources are difficult to tackle using methods other than type (1). Reduction of eutrophication in rivers relies on a reduction in inputs by campaigns at the practical level, such as cutting the use of fertilizers and bad spraying practices or by installation of good treatment plants, diversion, storage and settlement of polluted water. On short stretches, environmental manipulation can be used. Structures such as weirs can increase oxygenation and bubbler machines have been used, for example on the River Thames.

Lakes are more amenable to all techniques and considerable success has been achieved in many cases. A whole series of trials have been run in the Norfolk Broads, notably at Alderfen and Cockshoot (Moss, *et al.*, 1986) and a Royal Society for the Protection of Birds reserve at Strumpshaw. The inspiration lay in the observation that the few remaining undegraded broads were often isolated from the adjacent rivers, coupled with Swedish experience of nutrient treatment and active vegetation and sediment removal, with unique management programmes designed for individual lakes (Björk, 1972). Attempts have been made to remove nutrients from the river waters directly, by abstraction and stripping, but loadings remained heavy and costs high. At all three Broads isolation was carried out by dams and sluices across inflow dikes, and at Alderfen by additional diversion of an inflow stream with effluent from a small sewage plant. At Strumpshaw, improved embankments isolated the fen and wetlands from general winter flooding. Strumpshaw's recovery has been breathtaking. Following isolation in 1978 seventeen macrophyte species were recorded in the dike behind the dams, mostly natural re-establishment. Frogs, toads and fish spawned. The dam and new embankments were reinforced in 1980, old dikes redug and a new Broad excavated by dragline. The old Broad, which was completely filled in with mud and vegetation was cleared out. Floating water jets, towed back and forth, broke up the vegetation and mud. Suction pumps then removed the sediment to dumping sites that would not drain back into the Broad. From 1980 to 1986 some topping-up with nutrient-rich river water was allowed to manage the wetland habitat but no detrimental effects were noticed. The final coup was to install a borehole so that no river water need be used and a clean, controlled supply maintained. At the time of writing, this has been operating for over two years but by a nasty irony the borehole supply may not be clean and a special water quality monitoring project has been set up.

The restorations of Cockshoot and Alderfen have been carefully monitored (Moss *et al.*, 1986). Prior to isolation (pre-1980) and for the season afterwards Alderfen had high phosphorus (total P peaks of 900 μg^{-1}, Soluble Reactive P (SRP) peaks of 130 μg^{-1}) and dense summer phytoplankton with chlorophyll a at midsummer over 200 $\mu g\ l^{-1}$ dominated by a colonial blue–green alga. Macrophytes have been almost completely absent since the 1970s. From 1980 to 1982 the situation improved. Phosphorus levels fell and remained roughly constant. In 1981 total phosphorus peaked at 120 μg and SRP at 50 $\mu g\ l^{-1}$. Phytoplankton blooms were limited with chlorophyll a less than 50 $\mu g\ l^{-1}$, dominated by flagellates and green algae. Hornwort, *Ceratophyllum demersum,* covered 10 per cent of the bottom and in 1982 formed thick swards. However, from 1983 to 1985 the system declined again. Total phosphorus peaks exceeded 900 $\mu g\ l^{-1}$ with wide variation, SRP reached 700 $\mu g\ l^{-1}$ with wide variation, SRP reached 700 $\mu g\ l^{-1}$. Hornwort growth crashed and turbidity increased with the blue–green alga *Anabaena* dominant. In this example the progress from high nutrient supply and phytoplankton blooms to macrophyte dominance had reached a third state, with high phosphorus release from sediments perhaps encouraged by weed bed conditions, rekindling a eutrophic state. For, in this instance, unlike Strumpshaw, the nutrient-laden sediment had not been removed.

At Cockshoot, the Broad had all but filled in by 1980 and was pumped out to a depth of 1 m over 75 per cent of its area. Total P peaks exceeded 300 $\mu g\ l^{-1}$, SRP 50 $\mu g\ l^{-1}$, chlorophyll a in summer was over 200 $\mu g\ l^{-1}$, dominated by the blue–green *Oscillatoria.* By 1985 phosphorus levels had steadily declined and stabilized with 1985 total phosphorus reaching 100 μg

and SRP 10 $\mu g\ l^{-1}$, Chlorophyll a peaks fell below 50 $\mu g\ l^{-1}$ dominated by small flagellates. The phytoplankton changes may have been due to increases in *Daphnia* populations, though these zooplankters are now suffering from fish predation as fish recolonize. Thirteen macrophyte species have re-established, with the help of artificial introductions. As at Strumpshaw, isolation and sediment removal appear to have worked. In larger, deeper systems aeration of deep water and sediments has been used. This can be achieved by destratification, pumping hypolimnial water to the surface to oxygenate, then back down again, or by aeration of the hypolimnion with a submerged pump (Jorgensen, 1980; see also Fig. 7.6).

The plant introductions and *Daphnia*–fish interactions at Cockshoot are aspects that can be developed further as biomanipulation. Biomanipulation experiments have involved reduction of planktivorous fish populations, by netting, poisoning or introduction of piscivorous fish. The intention is to promote grazing zooplankton recovery. Shapiro and Wright (1984) describe such an experiment which did allow *Daphnia* recovery in a natural lake and apparently led to phytoplankton decline. The introduction of plants may aid restoration to a non-eutrophic state. Even the use of artificial cover may provide sufficient protection for the recovery of zooplankton.

The message of these small-scale restorations is optimistic. We can restore eutrophic waters. For larger systems physical intervention is difficult. Reduction of inputs is currently the only practical treatment.

7.6 Hypertrophic systems

Eutrophication is a recognizable suite of processes and symptoms but there is a more extreme condition that has been described as more than simply eutrophic (Barica and Mur, 1980) and dubbed hypertrophic, or sometimes hypereutrophic (Fig. 7.7).

These waters are characterized by very unstable oxygen levels, regular fish kills, noxious algae and bacterial blooms. There are extreme fluctuations of water quality and productivity on a seasonal, even daily basis. Examples include lakes with high and uncontrollable nutrient input, fish ponds and sewage purification lagoons.

Eutrophic conditions, degraded though they are, tend to have a predictability about them. Hypertrophic waters seem to represent a lurch into an unstable, chaotic regime. Certain features are noteworthy. Oscillations in conditions are erratic and extreme. Oxygen levels can often be very high, with 200 per cent saturation, at which levels oxygen itself is dangerous to fish. Nutrients are not exhausted even at maximum plant growth, due to the huge supply, in contrast to eutrophic systems where detectable levels may be very low as plants use all that are available. The lack of nutrient limitation frees the phytoplankton from many restraints and blooms tend to be continuous, rather than crashing. Blue–green algae are

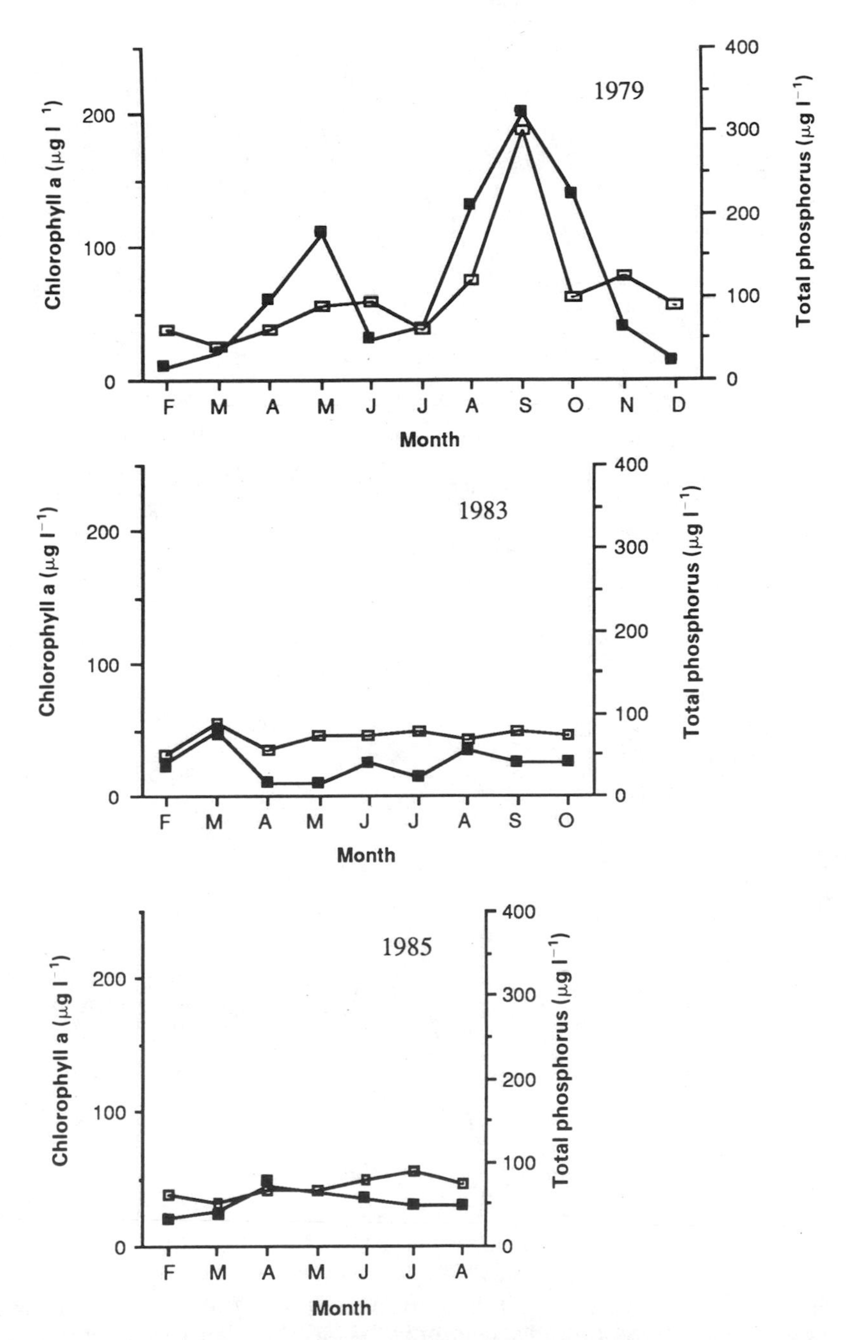

Fig. 7.6 (a) Effect of restoration on phosphorus (□) and phytoplankton, as measured by concentration of chlorophyll a, (■) of Cockshoot Broad. The restoration started after 1979. Total phosphorus and phytoplankton levels have fallen and remain at low, generally stable levels. (After Moss *et al.*, 1986)

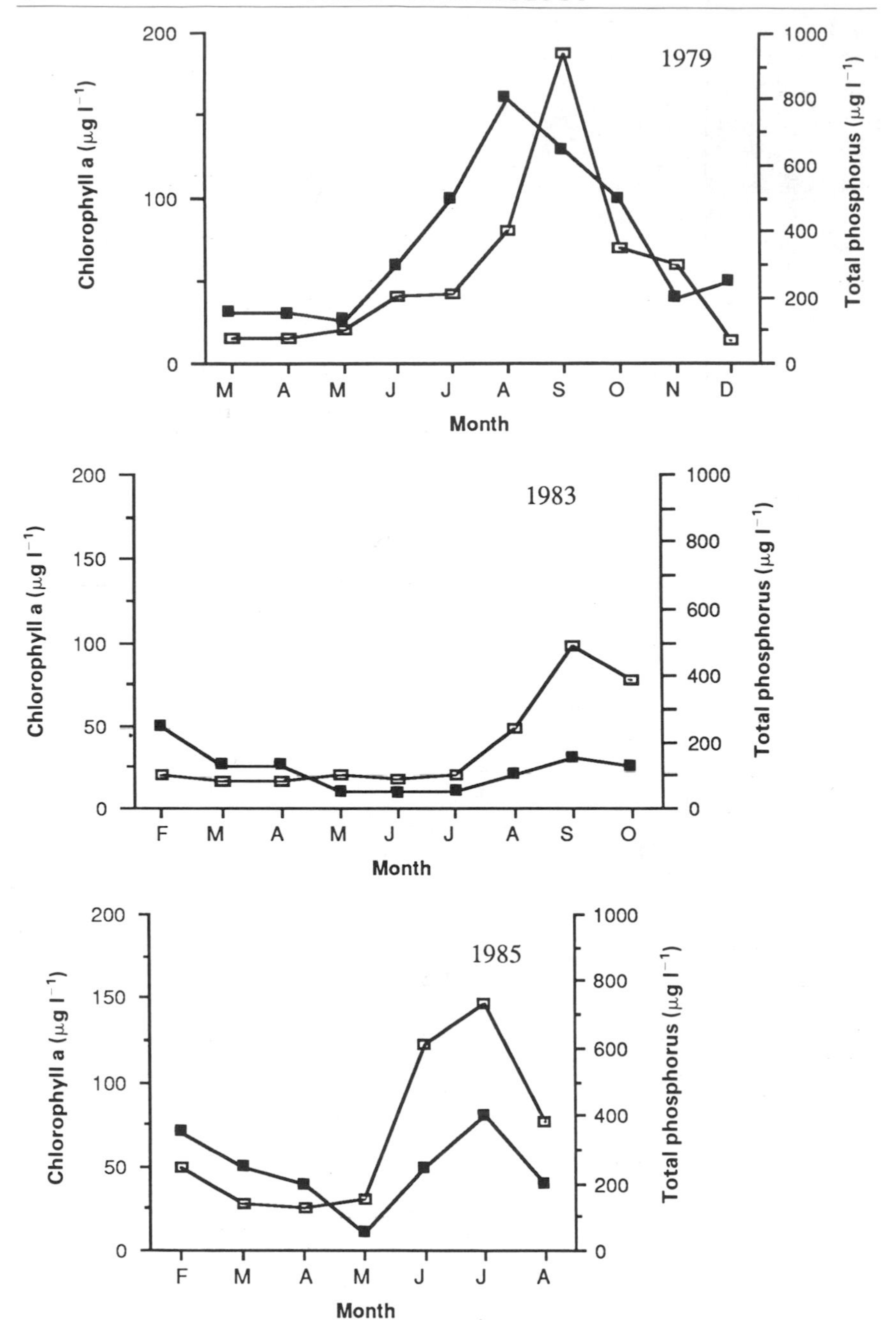

Fig. 7.6 (b) Effects of restoration on phosphorus, (□) and phytoplankton, measured as concentration of chlorophyll a, (■) abundance in Alderfen Broad. The restoration resulted in a decline in phytoplankton and total phosphorus but by 1983 phosphorus release from the sediments, which had not been removed, had re-established a high nutrient load, phytoplankton bloom ecosystem. (After Moss *et al.*, 1986)

frequently abundant but populations fluctuate sometimes on an hourly scale as conditions change. Overall productivity of dominant species is massive. There is a marked sensitivity to the daily changes and initial physico-chemical conditions (Fig. 7.8).

Hypertrophic waters may be just an extreme form of eutrophication, another qualitative switch in the community and habitat just as cultural eutrophication is from the natural nutrient cycle. However, their characteristic chaotic, immensely productive regimes are worthy of note as a most unusual form of ecosystem.

7.7 Organic pollution and saprobic systems

Another unexpected feature of hypertrophic systems is that, even if oxygen has been supersaturated during daylight, there may not be the expected oxygen sag at night. This all depends on the organic content and associated microbial activity. Pollution by nitrogen and phosphorus nutrients and pollution by organic material are not one and the same, even though in practice the two are commonly associated. General organic pollution represents the commonest form of general degradation in freshwaters.

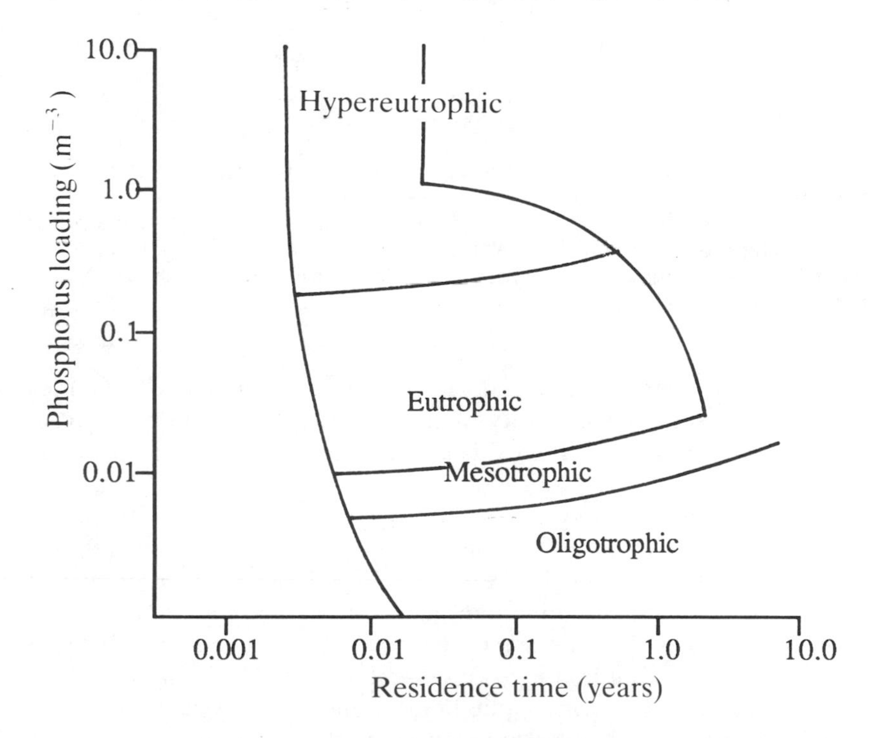

Fig. 7.7 Trophic state as defined by residence time of water in a lake system and phosphorus concentration. (After Jorgensen, 1980)

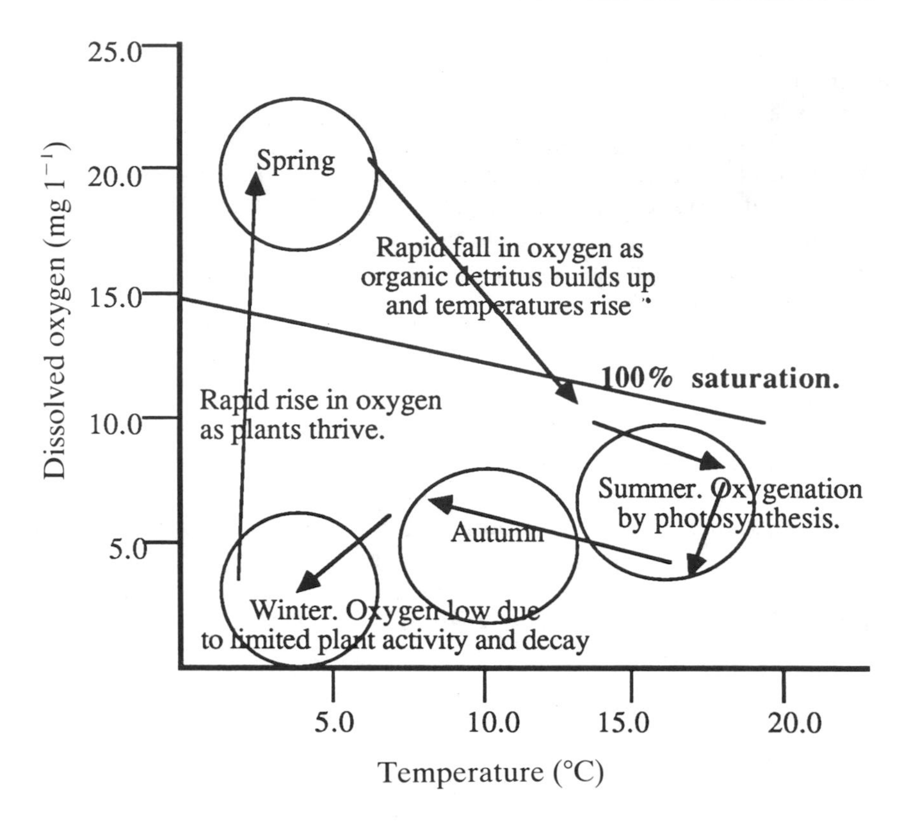

Fig. 7.8 Seasonal temperature–oxygen regimes in a hypertrophic system. Conditions tend to be steady within a season then switch rapidly to another state. Note especially the spring oxygenation and dramatic fall to summer regime as temperature and decomposition affect available oxygen. The 100 per cent saturation line of oxygen for the ambient temperature is shown. (After Barica and Mur, 1980)

The amount of organic matter and the activity by microbial communities living on it is called the saprobity of the ecosystem. Saprobity is a term introduced in Germany early in this century for the assessment of water quality, and the use of saprobity as both a term and practical approach has been primarily in Europe. Waters are said to have a saprobic level, which can be measured using the species present and their relative abundance, in effect a biotic index of organic pollution. The communities change, both qualitatively and quantitatively as organic content increases. This change is a saprobic succession, an ecological succession just like that of macrophyte communities in a lake, except involving bacteria, fungi and protozoa. Saprobity has been applied more widely to include not only the microbial decomposition but also respiration by all primary consumers, a measure of the total respiration of the ecosystem, but the term is still primarily associated with organic content, microbial activity and succession. The communities have been described and classified to provide an index of saprobity. Sladecek (1979) provides a review of saprobic systems, integrating this approach with other measures of water quality.

Organic pollution arises from many sources, although inputs from sewage, agriculture and some industrial processes such as brewing and paper making have been particularly widespread. The characteristic pattern resulting from the increase in organic material is oxygen depletion and microbes flourish on the detritus. The microbial activity is commonly measured by the biological oxygen demand (BOD). Immediately downstream of the effluent's source a rapid fall in oxygen, called the oxygen sag, occurs, followed by a gradual recovery, as the water re-aerates and material is used up. This is called self-purification. While the oxygen depletion is the main cause of degradation, the organic effluent is typically turbid, with a high suspended solids load, and this alters the natural sediment. The effluent may also contain toxins. The saprobity system distinguishes four phases. The polysaprobic system describes the severely degraded state just downstream of the discharge, with anoxic conditions not uncommon. The α-mesosaprobic stage represents the start of recovery with anoxic mud and noxious smells no longer dominant, though animal life is still restricted, the β-mesosaprobic stage following on with marked faunal recovery and declines in bacterial numbers. The oligosaprobic stage is that in which a full recovery is made, either to the same community that lived upstream of the pollution or to the appropriate, natural community for the river channel as it stands downstream. The microbial community changes are especially dramatic and sewage fungus blooms may develop. Sewage fungus is really a whole community of over 100 different bacteria, fungi, plants and animals of which the bacterium *Sphaerotilus natans* Kutzing is often dominant. The fungi and bacteria form slimy tufts and wefts, smothering substrate, plants and animals. Heavy growths are associated with a BOD of 5.0–30.0 mg $O_2 l^{-1}$ and soluble organic carbon of 6.0–30.0 mg l^{-1}. The many microbes each have different tolerances and it is the microbial wildlife that is the basis of saprobic indices of organic pollution. Macrophytes are generally severely affected by turbidity, abrasion from the suspended solids, sedimentation and smothering.

Macroinvertebrate responses have been extensively studied. A very few species can tolerate the polysaprobic state, notably air-breathing rat-tailed maggots (*Eristalis* spp.) and oligochaete worms of the genera *Tubifex*. Numbers of worms can be enormous as the organic debris and microbes provide a rich food source and competitors and predators are excluded by the grim conditions. Midge larvae of the genus *Chironomus* are also typical of these conditions, their bright red bodies packed with haemoglobin to glean the most of the limited oxygen from the water. In the mesosaprobic recovery stages a sequence of other invertebrates appears. Firstly *Asellus aquaticus* L. and various tolerant leech and snail species, then *Gammarus pulex* and tolerant mayflies (e.g. *Baetis* spp.) and caddis. The most sensitive groups include many mayflies (e.g. *Ecdyonurus, Ephemera*) and stoneflies (e.g. *Perla, Isoperla*) which are very intolerant of low oxygen. A saprobic classification for invertebrates has been compiled by Hellawell (1986), including examples from all around the world. Fish are also affected by low oxygen, suspended solids and toxins and responses range from brief emigration to complete absence. Very thorough reviews can be found in James and Evison (1979) and Hellawell (1986).

Organic pollution from domestic effluent is now much less severe in many developed countries with effective pre-discharge treatment. However, there is still a problem from agricultural effluents and some new sources have arisen, such as fish farm wastes.

8 Acidification of freshwaters

The acidification of freshwaters is part of a wider environmental problem resulting from the deposition of acidifying chemicals into many habitats. The problem is not restricted to aquatic ecosystems but they have proven especially vulnerable, often the first to show dramatic changes and attracted the first research into acidification. The problem is not quite the global phenomenon that eutrophication has become, with most polluting sources and damaged areas restricted to Northern Hemisphere industrialized countries, but it is naive to imagine the consequences are not potentially worldwide. In common with eutrophication, acidification of freshwaters comes about by the perturbation of vitally important natural chemical cycles. These operate on an international scale, with polluting emissions from one country falling on another. Acidification has also proven a complex tangle of linked factors, with not only the pollutants but also land use, climate and natural geology all complicating the results. For all the apparent simplicity of the idea of 'acid rain', unravelling the causes, consequences and likely worth of any cures has not been at all easy.

8.1 Natural acidity

Firstly, a reminder that some habitats are naturally acidic and support a valuable wildlife of their own. Unpolluted rain is a weak acid, carbonic acid, that readily dissociates to free H^+ ions, which increase acidity, or reassociate, mopping up excess H^+ ions so taking acidity out of the system, these reactions depending on the natural acidity of the landscape. For most waters this creates a buffering capacity against extremes of acidity. The final pH of a water body depends on the surrounding catchment, especially the geology and its ability to provide basic ions, and the vegetation, especially *Sphagnum* moss species which enhance acidification. Aquatic vegetation can be important as a source of OH^- ions, released during intense photosynthesis, which will alter pH perhaps on a daily cycle. Sometimes it is possible to distinguish the gross cause of acidification, be it natural, from changes in the catchment (natural or due to humans) or from anthropogenic pollution. Batterbee *et al.* (1985) analysed algal and pollen remains from an acidified loch in south-west Scotland. Acidification may be due to atmospheric inputs or land use changes resulting in regeneration of acidic heath in the catchment. Pollen analysis rebutted the suggestion of heathland regeneration as the cause, with no apparent increase in heath species' pollen. The change appeared to be anthropogenic. The general result of these buffering and catchment interactions is that most waters are circumneutral but a few systems are naturally acidic and provide a reference against which to gauge the impact of pollution.

8.1.1 Acid bogs

The most extensive acid freshwaters are the tracts of blanket bog found in both Southern and Northern Hemispheres beyond 40° north or south. Blanket bogs in Australia and Scotland will differ greatly in detail, such as species, but have common features of peat, bog structure and topography. Bogs have extensive peat, formed by organic accumulation, notably *Sphagnum* moss species. They can form on flat ground, in basins, even on slopes up to some 30°. Climate is the crucial factor with a high, even rainfall and low temperatures required. General global criteria for bog formation are a minimum annual rainfall of 1000 mm, a minimum of 160 wet days a year and a cool climate with mean temperatures of less than 15 °C for the warmest month (Ratcliffe and Oswald, 1988). Note that these are the conditions for active growth, though bogs will remain, inactive, if conditions change. The high rainfall will leach chemicals from the surface soil, commonly redepositing them lower in the profile to form an 'ironpan'. This characteristic soil, leached above with an ironpan below is called a podsol. Ironpans are largely impermeable, causing waterlogging. Impermeable geology and acidic rocks, or rocks that weather too slowly to buffer effectively, are also associated with bog habitat. In the base-poor, waterlogged landscape *Sphagnum* species thrive, further enhancing acidification by binding up cations and releasing H^+ ions in their place. Acidity, waterlogging and anaerobic conditions increasingly inhibit natural decay. The organic build-up will release organic acids to add to the problem. In small basins these processes result in raised bogs, often the final stage of a hydrosere succession even in areas of generally rich geology (e.g. Askham bog in Yorkshire). The encroachment of *Sphagnum* across other terrestrial vegetation, even woodland, with the waterlogging and peat production is paludification and results in extensive flat blanket bog. This is typically strewn with intricate pools, hollows and channels, called the bog microforms and the precise patterns vary with geography and topography. The water is peat stained and the sides are steep with sediment and unstable peat debris. Overall then, peat bogs are waterlogged, acidified and covered with small lentic habitats. The Flow Country blanket bogs in the north-east tip of Scotland are a superb example of the existence of a typical bog flora and fauna. Different communities of plants develop varying slightly with the microforms. Several *Sphagnum* species straggle through the open water. In shallow pools, cottongrass, *Eriophorum angustifolium* Honckeney and bogbean, *Menyanthes trifoliata,* are emergents (Ratcliffe and Oswald, 1988). Animal communities are also impoverished, dominated by dragonflies, bugs and beetles. They are top heavy with predators, perhaps because the primary production and detritus food trophic links have been destroyed by the acidity (Fig 8.1).

Anthropogenic acid deposition does affect these habitats, not so much by dramatically tilting the acidity balance but by increased stress from nitrogen and sulphur, which also affect the nutrient cycles. The stress may make the habitats vulnerable to dangers such as erosion.

8.1.2 Lake systems

Some lakes are naturally acidified. There are two main forms. Ultra-oligotrophic waters, in areas stripped by glacial action and on barren resistant geology may have very little buffering capacity, and typical pH of 5.5–6.5. This is on the acidic side, though not severe. Such lakes perhaps count as very vulnerable to anthropogenic inputs rather than acidic.

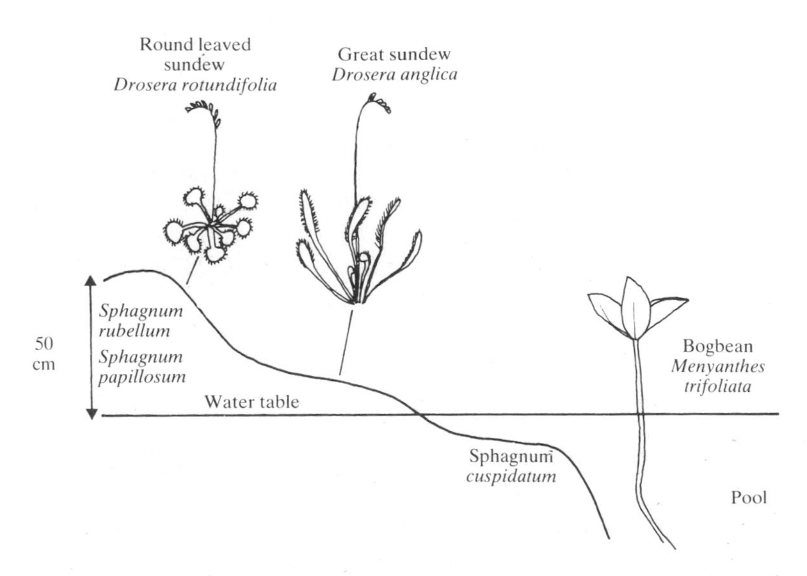

Fig. 8.1 Vegetation of acidic blanket bog. A cross-section of bog topography is shown, with water table and associated plant zonation. This is a typical general pattern on British bogs. (After Ratcliffe and Oswald, 1988)

The second form is lakes, again associated with acidic catchments, but often surrounded by forest, heath or bog. They are typically deeply peat stained, with waters of pH 4.0–5.5, representing dystrophic conditions, where organic matter is plentiful (in contrast to the ultraoligotrohpic lakes) but acidity has prevented normal decay and cycling. Rask *et al.* (1986) describe the ecology of such a lake in Scandinavia where they are associated with forest. Similar sites also occur in Britain. The Uath Lochans in Speyside, Scotland, are a fine example of how different acidified and circumneutral waters can be (Table 8.1). Four lochans occur within a few hundred metres of each other. All four appear to be depressions, perhaps kettle holes, left in morainic deposits as glaciers retreated. Two are circumneutral, pH 5.5–6.9, oligotrophic, gently shelving and supporting diverse animal and plant communities. The other two are steep sided, dystrophic with pH as low as 4.1 (Charter, 1988), with only a few species of plentiful numbers of beetles, bugs and dragonflies, one species of aquatic macrophyte, the white water lily, *Nymphaea alba,* and rings of *Sphagnum* around their edges. Since all four are so close together, apparently on the same glacial soil, surrounded by forests and heath, pollution differences are an unlikely cause. Instead two have noticeably acidified, perhaps due to inputs from *Sphagnum* and heath which could not be buffered by the steep, deep waters lacking macrophytes. As with eutrophication, perhaps there is a delicate balance between acidified and unacidified conditions. Should the balance tilt slightly one way the process accelerates and dystrophy and *Sphagnum* encroachment become self-reinforcing.

8.1.3 Lotic habitats

The small streams draining acid habitats will also be acidic but naturally acidified larger rivers do not occur. Moving waters do not accumulate peat or allow the self-sufficient cycle of acidification to start. However, as rivers move through a catchment, acidity will vary with surrounding geology, soils and land use between stretches. Townsend *et al.* (1983) describe a fine example from southern England, where geology and human activity normally conspire to give eutrophic, base-rich waters. Two rivers, the Medway and Sussex Ouse were studied. Both have catchments on sandy, acidic southern heathland. The soils contain little buffering capacity and ironstone oxidation plus *Sphagnum* mires contribute to acidity. In some headwater streams mean annual pH is below 5.0 downstream, in areas enriched by agriculture mean annual pH ranges from 5.8 to 6.9. The relationships of animal community to physico-chemical environment were analysed in many ways (also see Chapter 5). Total numbers of individuals, of taxa and of species diversity all increase with rising pH. Numbers of taxa and individuals in all trophic categories, except shredders, significantly increase with pH. Shredder numbers decrease with pH. Finally, the analyses

Table 8.1 Acidification of vulnerable lakes. The Uath Lochans, Glen Feshie, Scotland. These four small lochs are all within a few hundred metres of each other and partially linked by drainage. They all lie on the same glacial deposits and are surrounded by conifer forest and heather moor. Two (1 and 3) have apparently acidified, with limited plant communities. This may be a natural process as 1 and 3 are steep sided and once acidification started the influence of *Sphagnum* and build-up of peat sediments have encouraged dystrophy and their current species-poor wildlife. Lochans 2 and 4 have not acidified. (After Charter, 1988 and authors' data)

	Lochan number			
Character	1	2	3	4
pH	4.8–6.4	5.5–6.8	4.1–4.8	5.7–6.9
Alkalinity ($caCO_3$, mg l^{-1})	3.5	7.0	0	10.0
Substrate	Peat	Peat, pebbles cobbles	Peat	Peat, gravel, cobbles, pebbles
Main vegetation communities, from NCC classification	*Carex rostrata*	*C. rostrata*/bogbean/horsetail and common reed/bogbean	*C. rostrata*	*C. rostrata*/bogbean/horsetail and common reed/bogbean
Extent of *Sphagnum*	Floating ring around all of lochan	Limited to a few patches	Floating ring around all of lochan	Apparently none around waterside
Molluscs	None	*Lymnaea peregra*	None	*Lymnaea peregra*

were taken to species level. Some, such as *Gammarus pulex*, showed positive correlations with high pH. Others such as the stonefly *Leuctra nigra* (Olivier) showed a strong negative correlation with pH, that is there were more of them at lower pH. These stonefly nymphs may be feeding on iron bacteria, present where high iron concentrations alter pH (Table 8.2).

8.1.4 Geothermal habitats

The final group of natural acid waters are those created by geothermal springs. Geothermal waters may be very acid or alkaline. The bizarre and ferocious chemistry results from the heat, pressure and depth from which they arise. Acidic examples typically have pH ranges of 2.0–3.0. An extraordinary example is quoted by Bayly and Williams (1973) of a crater lake on Mount Ruapehu in New Zealand with a measured pH of 0.9. Investigations of these systems involve great risks. Falling in can prove rapidly fatal and the waters are so acid that they dissolve equipment. In less extreme cases, especially streams flowing away from geothermal springs, neat successions of plant and animal communities develop, reflecting the progressive cooling, with different thermophilic algae and bacteria being dominant. Blue–green algae are noticeable absentees from such acidic waters. Their primitive form (they are closer to bacteria than algae) lacking many intracellular organelles, makes their metabolisms vulnerable to acidity (Fig. 8.2).

Table 8.2 The influence of acidity in a river system. In this study, by Townsend *et al.* (1983), the abundance of species of river invertebrates was analysed in relation to changes in selected physico-chemical factors. Certain species showed significant relationships with several factors. The results given here, for *G. pulex* and the stonefly *L. nigra*, show the factors that were significant in explaining the variation in abundance of each species. The cumulative variance is a measure of how much of the variation is explained by each additional factor, starting with the most important, in this case pH. *G. pulex* is positively correlated with pH, that is more abundant at higher pH. *L. nigra* is negatively correlated with pH, that is more abundant at acidic sites. *L. nigra's* preference for acidic conditions may be related to iron bacteria as a food source, note the positive correlation with total iron

Gammarus pulex			**Leutra nigra**		
Variable	Cumulative variance	Correlation	Variable	Cumulative variance	Correlation
pH	49.3	Positive	pH	65.5	Negative
Calcium	56.2	Positive	July temp.	79.8	Negative
Nitrate	60.1	Positive	Total iron	85.5	Positive
			Max. discharge	87.8	Negative

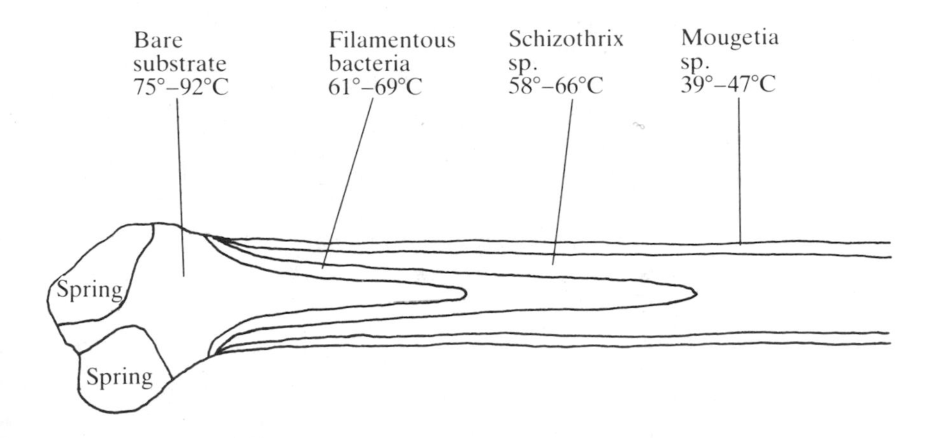

Fig. 8.2 Typical thermal spring and outflow stream zonation of algae and bacteria. Note in this example the water is alkaline and can support blue–green algae. The maximum temperature tolerances of each species determine how far up towards the spring they can extend. The downstream spread of each species is curtailed where the next most tolerant species is able to survive the physical conditions and out-competes the upstream species, generally by smothering. (After Stockner, 1967)

8.2 Anthropogenic acidification

Natural inputs into the atmosphere from sources as diverse as oceanic aerosols, volcanoes, planktonic metabolites and lightning will dissolve in water vapour and fall in rain. The chemical interactions of precipitation and soil determine the catchment's natural potential to acidify. Mankind has upset these processes in two main ways. Firstly, we have added to the atmosphere's chemical cocktail with gaseous emissions. Secondly, we have altered catchments through changes in land use. The suspicion that industrial atmospheric pollution may affect the environment has a long history (Cowling, 1980). Correspondence exists from the seventeenth and eighteenth centuries commenting on smoke emissions, and apparent poisoning around industrial works, even on the increased sulphur content of precipitation. Freshwater chemistry was only properly explored from the mid-twentieth century onwards by which time acidification was probably well established. The threat of acid rain, as it is now popularly known, was first fully investigated in Sweden in the 1960s and the growth of interest and research has been very sudden, in no small measure due to the dramatic damage the Scandinavians described.

The main culprit chemicals are sulphur dioxide (SO_2), hydrogen sulphide and various oxides of nitrogen. Sulphur dioxide is released by oil and coal burning, together over 50 per cent of emissions, plus many industrial smelting and extraction processes. Chemical pathways in the atmosphere are various but essentially all involve oxygen and water with the final product sulphur trioxide (SO_3), which dissolves in water vapour to give sulphuric acid ($SO_3 + H_2O \rightarrow H_2SO_4$). Note that pathways can involve either water, wet phase reactions, important in winter, or photochemical reactions with ozone (O_3) in bright sunlight, dry phase reactions that can predominate in summer. Either way the end result is sulphuric acid, a strong acid.

Oxides of nitrogen have an equally diverse chemistry, the more so since there are more types, sometimes collectively termed NO_X. Industrial combustion is a major source, but there are large natural fluxes due to the nitrogen cycle. Again reactions with ozone and water result in formation of a strong acid, nitric acid.

It is these solutions of sulphuric and nitric acids in rainwater that cause the classic acid rain, a term coined in 1872 by a Robert Angus Smith, Chief Alkali Inspector of the United Kingdom. Rain is not the sole source of inputs. Deposition can be divided first into wet and dry forms, the former involving water in some state, gas, liquid or solid, the latter not. Wet deposition further splits into precipitation or occult inputs. Precipitation involves rain, snow, sleet or hail actually falling on to a catchment. Occult deposition refers to mist or low cloud, hanging around a landscape and the mist aerosol settling out on surfaces such as leaves. Dry inputs can similarly be split into solid, particle deposition or dry aerosols intercepted by surfaces. Wellburn (1988) provides a very detailed and comprehensive review of the atmospheric processes and chemistry underlying acidification. Excellent summaries are given by Environmental Resources Limited (1983) and Mellanby (1988).

As a general consequence, rain falling on the eastern United States, north-west and central Europe now has an average pH of 4.0–5.0, with occasional events down to pH 3.5. The decrease from the expected natural pH of around 5.6 has been most noticeable since 1960. Remember that the pH scale is logarithmic, so that a one unit shift on the scale represents a real change of ten-fold in the acidity. This means the average rainfall in these areas is some ten times more acid than clean, natural precipitation. Occult deposition is important when cloud and mist contact the ground and vegetation surfaces. Many areas in Europe receiving atmospheric pollution have uplands with a wet climate prone to low cloud. The vegetation, especially extensive forests, acts as a colossal surface area on which droplets settle out. Forestry in uplands vulnerable to occult deposition may be a locally important source of acidified water. Acidity of droplets on leaves is increased by evaporation of water, or the equivalent for ice to vapour which is called sublimation, which concentrates the solution in the droplets or ice crystals left. Coniferous forests may be especially bad as needle litter alters soil chemistry and afforestation replaces other land use, such as pasture or small holdings which may have had a buffering effect.

Land use as it affects the catchment and eventually the lake or river has attracted increasing attention. Acidification depends on the interaction of the atmospheric inputs with soil chemistry. Since almost all land uses alter soil structure and chemistry one way or the other any land use has an effect on acidification. Simplistic generalizations such as afforestation causes acidification are naive, the consequences of change depend on the individual catchment. Once again, note how the whole catchment is an intimately linked unit.

Afforestation may well provoke an increase in acidity. The effects may be due to the trees directly increasing occult deposition. Conifers also release H^+ ions into surface soil horizons and relatively more humic acids than broadleaves. These changes alter pH and mobilize metals in the soil. The upper soil often dries out, with cracks altering drainage and promoting podsolization. Soil waters may penetrate less deeply and run off more quickly via the cracks. The deeper water penetrates and the longer it is retained the more buffered it tends to become. Ploughing prior to planting may decrease retention time and expose humic layers which leach acids. Deforestation can be harmful too. Clearance of native broadleaves can cause soil erosion, again leading to less retention and faster run-off. Acidifying vegetation such as heather, *Calluna vulgaris* (L.), heath may invade after clearance. Abandonment of pasture can be followed by heath invasion but animal manure itself is a source of nitrogen to soil. Moderate fires can damage soils and allow erosion. Hard fires, if they destroy significant amounts of humus may be beneficial with nitrogen and sulphur lost in the smoke, though presumably coming down elsewhere, and the ash left as a basic buffer. Agricultural practices such as liming will provide buffering capacity. These complex interactions of precipitation, land use and natural geology have resulted in a fragmented pattern of vulnerable, acidified lakes and others which have resisted damage (Batterbee, 1988). The diversity of processes is reviewed in Cresser and Edwards (1988) and Hornung (1988). The permutations are varied but all represent shifts in the balance of soil structure, chemistry and water. The detail of precise effects may be lost in broader patterns. Cresser and Edwards provide a good example from the Scottish Highlands. Acidification of rivers in this region may be partly due to an event in social history, the Highland Clearances. Many of the local population, living on small farms in the river valleys were driven off the land to make way for sheep. Whole communities and their associated small holdings were wiped out. As a consequence the mixed land use along the rivers, which included liming practices, was lost and often replaced by drainage and burning. In vulnerable catchments this in effect removed a buffer strip from along the rivers and increased acidifying land uses (Fig. 8.3).

To summarize, anthropogenic acidification is due to atmospheric pollutants deposited on catchments and interacting with vegetation and land use to alter soil and water chemistry. An additional complication is that severe acidification may only occur for brief periods after which tolerable

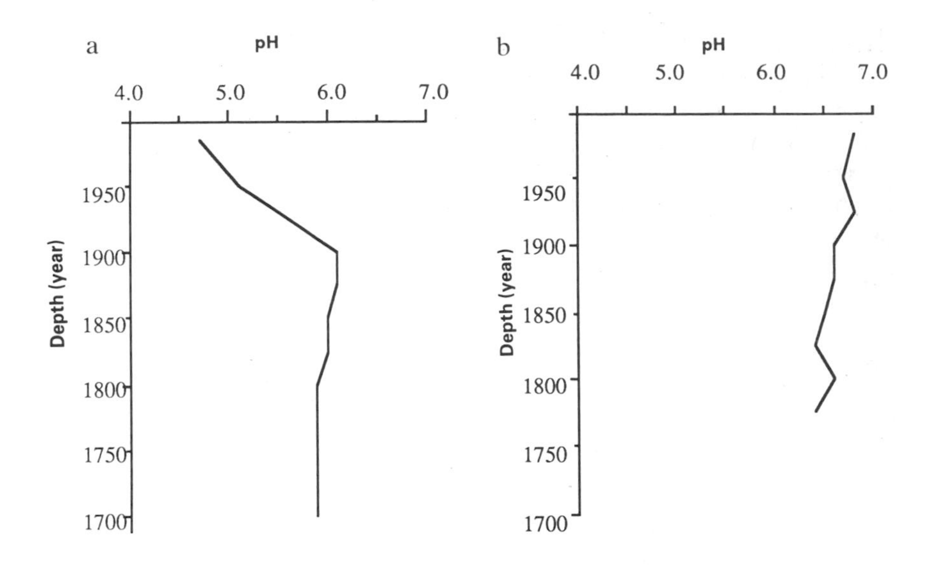

Fig. 8.3 The acidification of lakes based on lake sediment studies. The reconstructed acidification histories of two lakes (a) Llyn y Bi, Wales and (b) Loch Urr, Scotland, are shown. The reconstructions are based on studies on the diatom remains in the sediments. Different diatom species have different acidity tolerances and needs. The relative abundances of species can be used to trace changes in the waters' acidity. These two examples show (a) a lake that has undergone very abrupt change and (b) a lake on an area prone to acid rain but not itself on vulnerable geology and further buffered by catchment land use such as agricultural liming. (After Batterbee *et al.*, 1988)

levels may be restored, though the damage has been done. Such 'acid events' (or flushes, or shocks) are associated with bursts of high run-off or snow-melt. The surge of acidic water may kill or drive out the wildlife which remains depleted long after the event is over.

8.3 How acidification affects freshwater life

Just as the mechanisms by which water is acidified are many and various, so are the ways this harms the wildlife, ranging from the straightforward toxicity of acid through to complex changes in communities that have knock-on effects through foodwebs and competition.

8.3.1 Direct effects of H^+ ions

Hydrogen ions disrupt the ionic regulation of many organisms, which is of vital importance to life in freshwaters given the osmotic differences between organism and surrounding medium. Sodium and chloride loss can be severe, especially in the initial period, 0–12 hours following acidification. This shock effect may carry over and prove fatal, even if the acid event itself is brief. Influx mechanisms are disrupted but it is damage to exposed epithelia allowing loss from body fluids that seems the main cause. Concentrations which cause death are termed lethal, those which cause non-fatal injury and stress sublethal. Note that in the long term sublethal toxicity can still wipe out populations, because the surviving adults may not breed or younger stages may not survive to replace the natural mortality.

8.3.2 Elevated metal concentrations

Acidified waters dissolve out more metals and input from acid soil water can be a big source. Several very toxic metals such as mercury and zinc seem to increase in acid waters but it is aluminium that appears to be the main problem. Aluminium is an abundant element at the Earth's surface but much more soluble in acid conditions than circumneutral. Worse still, its toxicity increases at pH ranges typical of acidification. Changes in the toxicity of pollutants in response to other physico-chemical factors is a common aspect of pollution and these interactions are called synergisms. Aluminium exists in many forms. The most dangerous are labile, monomeric ions such as Al^{3+}, $Al(OH)^{2+}$ and $Al(SO_4)^{+}$. Non-ionic complexes also form, varying with the pH, in particular larger organic molecules. Aluminium interferes with ion regulation. Where H^{+} ion concentrations are causing damage though not fatal (sublethal) effects, the additional aluminium toxicity may be enough to finish the organism off. Gas exchange is also disrupted by interference with gill function. The irritation causes mucus secretion, clogging and eventual suffocation in fish. The larger complexes of aluminium can provoke this as well as ionic forms. However, since aluminium toxicity in the organically bound forms is minimal, rivers high in dissolved organic carbon, 5–15 $\mu g\ l^{-1}$ or more, are less at risk. Aluminium has also been associated with skeletal and developmental deformity, immunological and endocrinological disruption but the mechanisms are not understood.

8.3.3 Nutrient cycling

Aluminium forms complexes with phosphorus and may be responsible for taking phosphorus out of acidified systems. Since such waters are often suffering dystrophy, with normal decomposition and nutrient cycling disrupted, and commonly occur in areas with oligotrophic catchments this may be an additional burden.

8.3.4 Water clarity

Where metal complexes remove organic debris water clarity may greatly increase. Disruption of phytoplankton populations will add to this. It is an irony of many acidified lakes that they have enchanting, crystal clear waters, but this is a sterility rather than cleanliness. The increased light penetration allows acid-tolerant species to live deeper and new plant depth records have been recorded from acidified lakes in the United States (Singer *et al.*, 1983). Clear water will absorb and lose heat faster than turbid systems and another threat comes from thermal instability. Unusually rapid changes in thermal regime will be detrimental to many species.

The alternative effect of acidification in basins where there is a build-up of organic detritus is to cause dystrophy with its typical peat-stained water greatly reducing light penetration.

As a result of these physico-chemical changes the animal and plant communities will alter. These changes will themselves cause further disruption and the effects knock on through the ecosystem.

8.3.5 Decomposition

Changes in microbial communities alter decomposition rates. Generally processes are retarded and nutrient cycling decreased (Fig. 8.4).

8.3.6 Primary productivity

Losses of sensitive macrophytes and algae will lower autochthonous productivity. A frequent change in acid waters is for a smothering growth of algae to form a thick mat, often of just one species, on the substrate. This may be very productive but transfer of this productivity into foodwebs is limited as these acid-tolerant algae are often unsuitable food and poorly assimiliated by surviving grazers. Sublethal stress may also lower feeding by herbivores.

8.3.7 Trophic structure changes

One of the most complex changes is the alteration of foodweb structure, usually by loss of species causing a simplification and relative changes in the abundance of trophic categories such as shredders and grazers. In lentic systems, fish and sensitive zooplankton are lost. Large invertebrate predators take over. They are pH tolerant and also now free from fish predation. They in turn can act as keystone predators altering zooplankton populations and grazing pressure changes knock on in the phytoplankton.

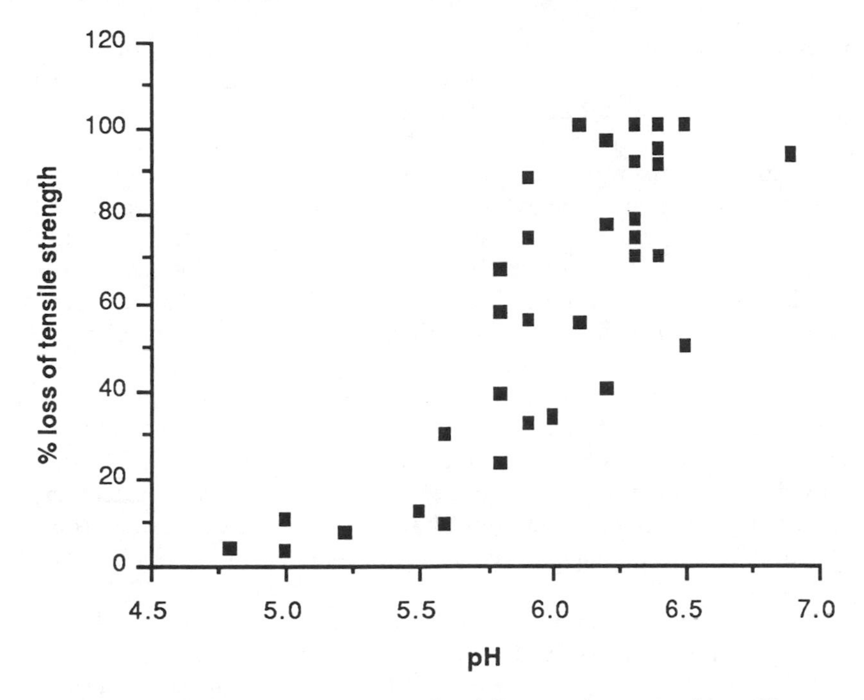

Fig. 8.4 Evidence for the impact of acidity on decomposition. The graph shows the loss of tensile strength of standardized cellulose cloth strips left in stream sites of various acidities. The strips were left for 49 days in winter. Losses of strength are lower at the lower pH, as acidity inhibits decomposition, so the strips are less damaged. In summer the effect of acidity is much less. Note that the effect is particularly severe at a pH of <5.5–5.8, the threshold often associated with marked damage to invertebrate communities. (After Hildrew *et al.*, 1984)

Loss of fish competitors has been credited with increased success in some visual hunting birds (Hunter *et al.*, 1986).

Trophic categories suffer differently. Grazers typically decrease as preferred algae are lost. Shredders may be unaffected, their relative importance increasing as grazers decline, even if their actual numbers do not.

8.4 Changes to the wildlife of acidified waters

Historically one of the first indications of acidification was the loss of fisheries from otherwise clean waters. Although this may be very sudden, due to severe acid events, the losses were usually the culmination of gradual change in the system with fish loss one of the last changes to occur. It was simply that it was the loss of fish that brought to our notice what was happening. Subtle alterations in other flora and fauna take place long before fish die out and studies have now looked at wildlife at all levels. It is worth appreciating the different approaches and their strengths and weaknesses used in these investigations. Studies of wildlife in acidified waters have relied on comparisons with apparently similar unacidified sites or comparison with historical records prior to degradation. Such techniques have expanded to include palaeolimnology using sediment subfossils as for eutrophication. This has allowed us to detect changes that happened long before we recognized acidification. Batterbee *et al.* (1988) provide an extremely detailed synopsis of acidification throughout Britain using these techniques. They detected changes in diatoms at sensitive sites with high acid deposition from 1850 onwards. Equally, little change was found at sensitive sites lacking acid inputs or non-sensitive sites even with acid deposition. The problem is that changes cannot be unequivocally assigned to acidification. Comparisons between different sites are also difficult since all catchments are idiosyncratic and the assumption that two lakes were the same prior to acidification and would have developed along similar lines if pollution had not happened is risky.

Field experiments acidifying a lake or stream, or perhaps parts thereof, are a direct approach akin to the manipulation techniques described for looking at competition and predation. This method has the advantage that the same water body can be sampled before and after acidification and, as long as the design and analysis are sound, changes can be confidently correlated with the acidification. However, subtleties may be missed since not every process can be monitored and gross, short-term modifications may not mimic the gradual changes that have occurred. Laboratory experiments can be used to measure the response of individual species to acidity changes and associated stresses such as aluminium. It is possible to work out tolerance limits which might be used to set standards for waters in the wild. What happens in a well-controlled laboratory and in the wild may be very different. No technique has a monopoly of perfection, but taken together and interpreted sensibly the general patterns of change and vulnerability have been thoroughly described.

Microbial activity is affected. Protozoa and bacterial numbers decline steadily from pH 7.0 to 5.0 and below 5.0 microbial activity falls sharply. Decomposer bacteria are replaced by fungi. Experiments measuring decay of leaf packs placed in streams show that decomposition slows, oxygen utilization is reduced and detritus accumulates. A neat way to try and standardize the detritus base used as a substrate has been to use manufactured cellulose sheets and use changes in strength to assess decomposition (Hildrew *et al.*, 1984). Algal communities alter. There is a change from planktonic dominance at pH 6.0–7.5 to periphyton at pH < 5.0. Dinoflagellates often become dominant in remaining plankton. Species composition changes with green and blue–green algae often predominating in thick mats at pH < 5.5. Acidity, metals and nutrient changes all contribute to this. Macrophytes have been rather poorly investigated. Again pH, metals and nutrients could all affect these plants. The *Isoetid/Littorella/Lobelia* flora typical of exposed oligotrophic lakes may be replaced by *Sphagnum*. Laboratory work has shown that increased metal concentrations depress oxygen production and delayed flowering in *Isoetes* and *Lobelia*.

Invertebrates' responses vary enormously and some groups have hardly been studied. The responses of molluscs, crustaceans, insects and zooplankton are best known. Molluscs need calcium for their shells. Gastropods and bivalves both show marked species losses below pH 6.0. The most tolerant, such as *Lymnaea peregra* (Muller) and *Ancylus fluviatilis* (Muller), live down to pH 5.5 and some *Pisidium* mussels can survive at pH 5.0. Crustaceans are also vulnerable due to calcium needs. The isopod *Asellus aquaticus* can survive to pH 5.0, the amphipod *Gammarus pulex* is more sensitive and will not survive below pH 5.5 and avoids waters of pH < 6.0. Crayfish, with their calcified shells ought to be very vulnerable and losses of crayfish, a delicacy in Sweden, were another incentive to initial research there. However, vulnerability is very variable with species. Some American crayfish live in lakes of pH 4.8–5.0. Generally pH < 5.5 is stressful, especially at the recalcifying stage after moulting. For all these taxa with high calcium requirements the measure of pH may not be the best criterion but some measure of calcium availability may also be needed.

Insects fall into three broad groups, generally intolerant, generally tolerant and orders with considerable interspecific variation. Mayflies (Ephemeroptera) are vulnerable, with species losses at pH < 6.5 and only a few species surviving at pH 4.5. Alternatively, beetles, bugs, dragonflies and damselflies are very tolerant, many surviving down to pH 4.5. Members of the caddis, stoneflies and alderflies (Megaloptera) vary greatly, with losses at pH < 5.5 but representatives down to 4.5. Taxa such as oligochaetes and leeches are not well researched but as a general rule diminish at pH < 5.0. Some of these losses may not be due to direct toxicity but as foodwebs collapse. Microhabitat is important. Species living in the sediment seem to be more resistant, as do those lacking external gills.

Plankton show the same declines as the benthos with losses at pH 5.0–5.5. The results are simplified communities impoverished by 40–80 per cent and taxa such as the waterfleas, *Bosmina*, and copepods, *Diaptomus*, replacing sensitive *Daphnia*. Populations are increasingly unstable with tolerant and sensitive species rapidly fluctuating if pH changes back and forth around this threshold.

Fish have been intensively studied in part to draw up effective guidelines to safeguard economically important stocks (Alabaster and Lloyd, 1980). Many species tolerate pH down to 5.0, especially if declines are gradual, giving time to acclimate. Below pH 5.0 ionic regulation damage and aluminium toxicity effects show. Salmonid eggs are vulnerable at this level. This differential effect on eggs and young may explain the typical populations of acid waters, which consist of large, aged fish (Maitland *et al.*, 1987). Lack of recruitment, as breeding fails repeatedly, will eventually result in the population dying out. Below pH 4.5 most adult fish suffer severely and the extreme limits for salmonids are pH 3.5–4.0. The rate at which pH falls and the ambient pH fish normally experience are important. A sudden, severe acid flush passing briefly through normally circumneutral waters may kill all the fish even if the pH has not been lowered to some of the laboratory derived extreme tolerances.

Vulnerability of other vertebrates is only patchily understood. Amphibia are susceptible as eggs and young. They seem intolerant of pH < 5.0 for successful breeding, except the palmate newt down to pH 4.0. Breeding success of birds varies. In Canada, fish and invertebrate feeding waterfowl such as the mergansers (*Mergus* spp.) and kingfisher *Alcedo atthis* (L.) only breed on lakes of pH > 5.6. Dipper success in Britain may be reduced by acidity and aluminium, as described in Chapter 5. Riparian habitat is a rich food source for insectivorous birds and their success may wane if fly hatches fail. Alternatively some may be more successful. Breeding by goldeneye *Bucephala clangula* (L.) on acidified clear waters appeared to rise, perhaps due to the loss of competing fish and also due to the increased visibility of large insect prey. Heavy metal accumulation might be an additional threat to birds and also to mammals such as the otter. The otter would be vulnerable as ecosystems collapsed and fish and crayfish food vanished.

There are many thorough summaries of the effects of acidification on wildlife (Environmental Resources Limited, 1983; Altshuller and Linthurst, 1986; Hammerton, 1988; Vangenechten, 1988).

8.5 Prevention and cure

Several strategies are used to treat acidification, each aimed at a different stage in the process: reduction of emissions, land use strategies and intervention cures.

As with eutrophication, stopping the supply of pollutants is necessary to prevent increasing damage and allow successful restoration of waters already degraded. The control of emissions is not just a technological problem but also one of economics as the costs involved will affect the industries responsible. Sulphur emissions can be tackled by use of low sulphur content fuels, sulphur removal in combustion chambers and removal from emissions in chimneys just before release, the latter commonly called scrubbing. The technology to cope with all three exists but costs can be high. Estimates of costs to fit new power stations with scrubbers have reached 15–20 per cent of total capital costs, fitting to old stations 30–50 per cent and overall a 10 per cent cost on generated, coal-fired, electricity (Environmental Resources Limited, 1983). Removal of sulphur in chambers remains expensive and not yet commercially developed. Changes in coals and oils used are also difficult, with increased costs and social consequences as suppliers are hit by changes (Davies, 1988). None the less, reductions in gross sulphur emissions have been agreed by many countries with international agreements on goals and timetables, for example the '30% club', a subgroup of signatories to the United Nations Economic Commission for Europe Convention on Long Range Transboundary Air Pollution, committed to reduce their emissions by 30 per cent from 1980 to 1993. Reduction of nitrogen oxides from fossil fuels is possible, even within combustion chambers, with 40–70 per cent cuts at economic cost levels. Car emissions remain a difficult problem.

Evidence for any amelioration is still scarce, partly because reductions have only recently started. On a local scale there is evidence of beneficial changes (Dillon *et al.,* 1986), showing lake recovery around an individual industrial plant that reduced output. On a wider scale, there is evidence of a decline in SO_2 deposition in Britain since 1970 (Harriman and Wells, 1985), this change, predating 'acid rain' awareness, perhaps due to changes in industry. Batterbee *et al.* (1988) report that there has been no recent deterioration in many of the lakes they have monitored and evidence of initial improvement in some of them.

Previous sections in this chapter make clear the importance of water, soil and land use interactions in acidification. Careful management of catchments can help prevent and ameliorate problems on the broad scale of whole catchment planning and smaller scale of practical management techniques used on a hillside or riverbank. It has been increasingly recognized that not all lakes will acidify and that there are vulnerable catchments. These tend to be in areas of high acid input and lacking natural buffering capacity. Identification of such sites is a first strategic step. Once a catchment's likely sensitivity is known then sensible land use practices can be employed, in particular limiting afforestation to least sensitive portions and employing careful forestry practices. Small-scale practices include ploughing and draining designed to prevent massive, sudden run-off, perhaps with riparian buffer strips to further increase retention time. Drainage water can also be diverted to sumps or pools on base-rich areas, even artificially created, before entering streams. Areas of the catchments including existing plantations can be treated directly by liming to increase buffering capacity. Some large-scale projects experimenting with different liming techniques in a catchment are being undertaken in Scotland on Loch Dee (Tervet and Harriman, 1988) and Loch Fleet (Howells, 1986).

Actual cures for acidified waters rely on various liming procedures, adopted widely in the north-eastern United States, Canada, Norway and Sweden. Liming is a catch-all term covering additions of basic materials to the water to alleviate and reverse acidification. Many materials have been tried such as limestone rock, chalk, lime ($Ca(OH)_2$), quicklime (CaO), shell sand and basic slag (Dickson, 1988; Mellanby, 1988). Their effectiveness varies depending on solubility, duration, availability, practical difficulties and cost of application. Some may be dangerous in themselves for example quicklime. Techniques have been extensively developed in Scandinavia with a national programme in Sweden, started in 1977, having limed over 4000 lakes. Sophisticated treatments have involved use of special ploughs to seed sediments and alarms to detect acid events and initiate buffering treatments when they are actually needed. As a general guide some 3 g $CaCO_3$ m^{-3} will reduce toxicity effects and 8 g m^{-3} restore good alkalinity depending on lake turnover time. However, the effects may not last very long. Estimates for Sweden in 1985 suggest that 50 per cent of treated lakes have reacidified (Nystrom and Hultberg, 1988) and the economic costs have cast doubt on the technique as anything other than a short-term remedy (Hammerton, 1988). In the long term integration of emission control, land use and direct intervention is needed.

The general impact of liming reverses the degrading changes due to acidification. Alkalinity, pH and calcium levels increase, metals decrease. Impact on phosphorus is variable. The restoration may improve nutrient cycling, but phosphorus will form complex compounds with calcium salts as well as heavy metals, and so may be reduced as an available form. Biological components are restored. Phytoplankton, zooplankton and benthos diversity recover and decomposition rates increase. Vegetation changes back from low diversity mats with isoeted communities restored. Fish survival and reproduction are restored. It is worth noting that a complete return to a pristine community is unlikely, the damage is repaired rather than magically restored whole. Excessive calcium deposits, especially of limestone or finer debris, may be unsuitable for wildlife typical of the nutrient-poor waters commonly subject to acidification and restoration. Other problems are subtler. Restoration of the bulk of a lake may still leave the littoral zone vulnerable to acidic flushes, as this will be the first zone to be hit and may be only a small part of the total lake volume. This zone may not recover and, since the littoral is so important for autochthonous productivity, this lingering damage may be disproportionately important. An added danger is that aluminium is most toxic between pH 5.2 and 5.5. Severely acidified waters may lie below this and restoration elevates pH into this critical range. Sediments may also become increasingly steeped in metals if they precipitate out.

Increasing use has been made of computer models to try and predict the vulnerability of catchments and likely success of remedial action. Barth (1988) provides a recent synopsis of the prospects for remedial work including extensive use of models. Restoration is not straightforward, but done carefully, hand in hand with other control measures, it does offer hope that we can rehabilitate many of the water bodies currently affected by acidification.

9 Pollution

The definition given of a pollutant in Chapter 7 is so broad because of the sheer diversity of things that can cause pollution. Although certain images such as oil spills or detergent froths spring to mind, many pollutants defy neat pigeonholing and work in many ways. It is more the end result, a degraded ecosystem, that unites pollutants. This chapter is intended to give a feel for the variety of pollutants, their sources and the effects on animal and plant communities. The examples chosen show how both direct and indirect mechanisms operate, just as for eutrophication and acidification. Only a few cases are described in detail but the same general principles show in all of them. In addition, the use of communities for monitoring pollution is described. Hellawell (1986) provides a thorough review of how pollutants affect wildlife.

9.1 Types of pollutants

The examples of pollution described here fall into the broad categories outlined below. The categories are not precise, objective divisions and are not mutually exclusive. One of the great difficulties with tackling pollution is that one source may discharge several types of pollutant, with different modes of interacting action.

9.1.1 Non-toxic pollutants

Pollutants in this category are not in themselves poisonous. The materials are often present naturally and are important in normal ecosystem function. Their destructive action derives from the excessive amounts present, causing imbalance in the ecosystem and often changes to other physico-chemical processes, such as oxygenation, which then prove fatal.

SUSPENDED SOLIDS

Many discharges contain high levels of particulate debris, generally classified together as suspended solids. They are inert but still destructive. They can physically abrade and scour the habitat. Increased turbidity may reduce light below levels needed to sustain photosynthesis. The substrate, even the animals and plants themselves, may become clogged resulting in flow reduction and suffocation. Finally, when the particles settle out they may cause massive alterations to natural substrate composition. The damage is an anthropogenic form of the natural high loads that occur during floods and landslides such as a peat slide (McCahon *et al.*, 1987). In many cases toxic materials may also be present. Sources include washings and discharges from extraction industries (e.g. coal, sand, gravel and china clay), settling lagoons (e.g. fly ash) and many agricultural industries; disturbance by engineering either to the channel, such as channel alteration (Brookes, 1986) or laying pipelines across or adjacent to the river in highway construction and gas or oil lines; forestry, especially at the planting or extraction phase as machinery disturbs the ground and drainage is altered; catchment changes such as impoundments altering flow or discharging to flush out sediment; many agro-industrial plants such as papermills.

NUTRIENTS AND ORGANIC DEBRIS

Both nutrients and organic enrichment were discussed in Chapter 7. Again both are not directly toxic but represent an imbalance, generally an excess, in natural cycles. Consequences, especially perturbation of oxygen regime, can be directly lethal. Severe organic pollution, such as silage, can be so noxious and fast-acting that it could be regarded as directly toxic.

9.1.2 Toxic pollutants

Pollutants in this category are toxic, either lethally or sublethally. Indirect degradation, through nutrient and oxygen perturbations and eventually collapse of functioning communities may also follow. This category includes many materials that have no normal place in natural systems, many being man-made.

TOXINS

This covers the host of directly poisonous substances: metals, dissolved ions, gases, acids and alkalis, complex organic compounds and radioactive material. Sources are as abundant as the chemicals are diverse with many coming from the general category of industrial effluent. Some particular links between source and pollutant are notable. Metals, especially heavy metals, are often leached from mineworkings, both active and abandoned, underground or opencast and from the associated slagheaps. The disused lead mines of mid-Wales are still a source of lead and zinc pollution in nearby rivers. Mine pollution is also often acidic. This form of pollution is not uncommon in wilderness areas far removed from major industrial plants. Dissolved ions such as sodium and chloride can be a local problem, associated with urban areas and where salt is used as a winter de-icer for roads. In some areas use of toxins by poachers, including substances such as cyanide, has resulted in massive poisoning of wildlife.

PESTICIDES

Pesticides is a general term including all chemicals such as insecticides and herbicides designed and used to kill animals and plants. As such they have the same effect as other toxins but a distinction is useful since they are specifically designed to kill wildlife and in many cases intentionally added to aquatic systems for management purposes. These applications may be of chemicals targeted at certain taxa and apparently of little danger to other life, for example the piscicide Rotenone, the reed-specific herbicide Dalapon and the arthropod destroying agent Fenitrothion. However, general pesticides such as DDT have also been used against aquatic stages of disease vector insects and there is still a temptation to use unsuitable pesticides in water with unforeseen consequences. Input from agricultural run-off and overspraying can be a problem. Pesticides applied to submerged structures, such as organo-tin compounds used as antifouling paint on boats, leach into the wider system. Other point sources include sheep dips and discharges from factories that treat domestic materials, such as furniture and carpets, against household pests. Muirhead-Thomsom (1987) provides a thorough review of pesticides in freshwaters.

OIL, PETROL AND DETERGENTS

The grotesque marine accidents that plague us do not have their equal in freshwaters but oil and petroleum pollutants are common. Typically they are run-off, leaks and discharges from industry, often small scale, but occasionally refinery effluent affects big rivers (Meynell, 1973). Run-off from urban areas and roads, slag heaps and waste tips is a routine burden. Occasionally effluent or sudden spillage into rivers has been enough for fires to break out, covering the water's surface. Many oils are toxic. Damage also results from smothering of animals such as mammals and birds, leading to fatal cooling and drowning and suffocation of submerged insects that need to penetrate the surface film to obtain air, the oil film obstructing this renewal. The eventual breakdown products of oils provide additional organic input. Examples include oil in streams and in ponds, experimental oiling manipulations, effects under ice and fish avoidance reactions to an oiled stretch.

Detergents are lumped together with oils more by association than effect. The main problems of phosphorus input (see Chapter 7) and unsightly floating foam froths have abated by changes in chemistry. Detergents are also toxic. A more specific danger arises in water treatment plants where the slippery detergents can make conditions very unsafe for workers.

9.1.3 Thermal

Many industrial processes use water as a coolant, which is then discharged often perfectly clean but much warmer. Electrical power generation is the major source (Langford, 1983).

Since all animals and plants have thermal tolerance limits, a discharge may be lethal if beyond the threshold for a species. Sublethal effects are common. Life histories may alter, especially length of time taken for growth, generally shortened, and hatching period, generally brought forward. Species that rely on seasonal cues to promote development will be at risk if natural temperature regimes are obscured by discharges. Species may be lost by emigration or avoidance. Behavioural changes may alter the balance of competition and predation. New species may invade and become established. In Britain, thermal discharges have been credited with sustaining isolated pockets of introduced tropical fish species such as the guppy (*Poecilia reticulata* Peters) in the St Helens Canal, near Liverpool in Britain, warmed by glass factory effluent (Lever, 1979).

The temperature regime will affect many other physico-chemical factors, particularly by lowering dissolved oxygen which may be further reduced by enhanced decomposition. Simultaneously elevated temperatures will increase the respiratory demands of wildlife and so the habitat can no longer support the species. Thermal problems will be worse if the discharge varies rapidly. Extremes may be tolerated, either by a different community becoming established or acclimation of the species present, but rapid fluctuations will not allow this.

Associated problems of power stations are kills from animals drawn into intakes (and generally filtered by 'trash' screens), supersaturation of discharged water with nitrogen leading to disequilibrium and a form of the bends in fish and use of anti-fouling chemicals to keep pipes clean. These chemicals are then discharged with the used water. Cold water discharges are not a common problem from industrial sources but the discharge of hypolimnion water from impoundments can cause thermal pollution downstream. Warm water discharges can be put to good use as a residential heating source, for agriculture and aquaculture.

9.1.4 Pathogens

Pathogens include viruses, bacteria, protozoa, fungi and metazoan animal parasites such as flukes and tapeworms, with the prime concern being the direct health risk to humans. Their presence is often associated with other polluting effluent, notably sewage. Some parasites are water-borne infective stages, free living or carried by aquatic vectors ingested when a target host eats the vector or eventually swimming out in search of a target of their own accord. Diseases can be very dangerous such as bilharzia, caused by a genus of trematode worms parasitic in humans, with an aquatic stage passed inside snails, or irritating, such as 'swimmers' itch' resulting from the, usually unsuccessful, attempts of infective stages of worm parasites of other animals to penetrate human skin. Treatment of dangerous waters, such as bilharzia sites, may involve potential pollutants, in this case molluscicides. In developed countries pathogen control is primarily a matter of adequate sewage treatment. Anthropogenic discharges containing pathogens that threaten wildlife are not a common problem though fish diseases may be released in aquaculture effluent.

9.1.5 Human, recreactional impact

Besides our noxious chemicals and other effluents humans themselves can be a very destructive presence. Sometimes this is obvious from the litter we leave behind. This can be the general flotsam of our activities but some examples such as anglers' lead weights are specifically associated with freshwaters and are a direct toxic danger, particularly to swans and other waterfowl. In addition, trampling, vehicles and boats damage habitat. Thirdly, the disturbance caused to certain animals, especially birds and mammals, by our mere presence can be thought of as a form of pollution. Disturbance may respond to the ideas of regulation and control associated with more straightforward examples of effluent. The added twist to this pollution is that it stems largely from the very great pleasure we gain from the countryside in which water plays a great part, whether it be a quiet walk or the 'messing about in boats' idyll of Ratty in *The Wind in the Willows*. The nicer a site is the more people are drawn there, so pressure can be disproportionately worse at some of the most beautiful areas, of which the Norfolk Broads are a prime example.

9.2 Defining and quantifying the effects of pollutants

There may be a multitude of pollutants but the general ways they act and the methodologies used to describe and quantify these effects are well established. This is largely due to the practical need to know the levels at which pollutants cause various degrees of damage.

A lot of effort has been put into quantifying the effects that varying concentrations of toxins have on animals and, to a lesser extent, on plants. These are toxicity measures and, just as with acidification, can be divided into lethal and sublethal effects. An additional division gives some sense of the speed and intensity at which damage is inflicted. Acute toxicity refers to damage, typically lethal, inflicted in a short period. Alternatively, chronic toxicity is that which drags on, perhaps only ever causing sublethal damage but perhaps lethal in the long term. The effects may be sublethal on an individual, such as preventing reproduction, but in the longer term this will still destroy the population as no recruitment occurs. In some cases chronic toxicity may give way to acute problems as successive doses or continual exposure cause a cumulative build-up, crossing a lethal threshold.

Measurements of toxicity levels rely on a standardized scheme to describe the various degrees of damage done by a certain concentration of toxin and the time taken. Measurements commonly use standard organisms, fish species such as the trout or ubiquitous invertebrates such as *Gammarus pulex* or *Daphnia pulex*, kept in laboratory cultures and subject to carefully controlled doses in carefully controlled conditions. The data collected may be survival times for individual animals at varying concentrations or, time taken for 50 per cent of the animals of a population to succumb. Both methods result in a measure of damage done as dose changes and are commonly presented as toxicity curves, showing mortality as dose changes. From such data generalized standards can be drawn up. The lethal concentration (LC), sometimes called lethal dose, is a measure of the concentration causing a specified level of mortality in a specified time period. For example, the lethal concentration that causes 50 per cent mortality in 24 hours, or that causes 75 per cent mortality in 1 hour would be abbreviated as LC_{50} 24 h and LC_{75} 1 h, respectively. A commonly used threshold is 50 per cent, and the concentration required to kill half the population in a specified time is the median lethal concentration. The time it takes for damage to be done is also a useful measure. Even very deadly toxins may not cause immediate damage and the time it takes for animals to react is called the threshold reaction time.

The safe concentration is that at which no discernible effects occur in the long term, which ought to include at least one generation to allow for impact on reproduction or young.

These various levels and thresholds can then be used to set standards necessary for water quality. There are problems applying the results of laboratory bioassays to complex wild communities. The controlled laboratory conditions lack the variety of other physico-chemical factors with which toxins interact but at least bioassays provide guidelines which are usually applied with circumspect interpretation of several studies. Typically, the standards set for water quality are based on acute toxicity measures multiplied by application factors of 0.1–0.01.

Pollutants may interact with each other, decreasing or increasing the overall toxic effects. The simplest interactions are additive. In such cases the effects of the combined concentrations of two toxins are the same as for the same concentration of one or the other by itself. If the toxic effects are greater than simply additive, so that the combination is more toxic than one or other by itself, then the interaction is called synergistic. If the combined effects are less than simply additive then the toxins are interfering with each other and the result is antagonistic. Interactions occur with natural physico-chemical factors. Temperature, oxygen, pH and water hardness are especially important. Temperature and oxygen interactions may work through the additional metabolic stress at levels outside of the optimum. Hardness and the closely linked alkalinity and pH probably act by direct changes to the toxins, notably metals, altering the precise ionic forms involved. The general result is that many toxins are less poisonous in hard waters than in soft waters and standards set for these pollutants are often subdivided into hard/soft water recommendations.

The damage pollutants cause will also vary from individual to individual and from population to population. Individual susceptibility will depend in part on the conditions the animal or plant is used to. Exposure to non-lethal concentrations of a toxin can result in development of a degree of resistance. This is called acclimation, or conditioning. The same process works with non-toxic, natural physico-chemical tolerances. For example, wildlife can acclimate to higher or lower levels of oxygen or temperature within their ultimate autecological limits. Individuals acclimated to an elevated or lowered range of a pollutant are more resistant to further changes in this toxin, as long as the changes are increases or decreases respectively. Note that an individual acclimated to higher levels than normal will be more vulnerable to lower levels than a normal, unacclimated individual. Short-term acclimation is physiological or behavioural. In the longer term, a genetic component may be added by selection of acclimated individuals and dominance by their offspring if continued pollution confers advantages on acclimation. The result will be a population adapted to the unusual conditions.

Besides toxicological effects the indirect complications, described for eutrophication and acidification, also cause damage with other pollutants. Animals may leave a polluted site by active emigration or drifting in the current. Recolonization may be poor with adult insects not ovipositing and natural migrations curtailed or diverted as animals will not cross polluted zones, or even isolated discharges, denying them access to clean waters beyond. This is a particular problem for salmonid fish as polluted river mouths can provoke avoidance reactions. Life history perturbations take a toll, especially if environmental changes are too erratic to allow acclimation. Sublethal poisoning can alter behaviour and reproduction. Some pollutants can cause gross changes in energy flow. For example, herbicides result in massive macrophyte loss. This will destroy the macrophyte–epiphyte–

grazer community as plant structure and epiphyton are lost, but cause a shift to a detritivore community using the debris.

9.3 Pollution examples

The studies cited here are chosen to provide clear examples of some of the major forms of pollution. Not only are the typical, gross symptoms evident but in many of the examples additional complications arise. At the same time the general patterns of degradation apply widely.

9.3.1 Suspended solids

Construction work is a very common source of non-toxic suspended solids. Immediate damage is inflicted but longer term degradation varies with the intensity and duration of the input. Barton (1977) gives an example caused by highway construction across a stream in Canada. Sample stations were established above and below the construction site and physico-chemical conditions, fish and benthic invertebrate populations sampled before, during and after work. This allowed different stages in construction, such as building, rechannelization and post-construction natural spates to be differentiated. Appreciable physico-chemical changes occurred only in the suspended solids load and sedimentation rate. Upstream of the work suspended solids ranged from 1 to 10 mg l^{-1} over the two-year study, with sedimentation rates of 0.01–0.1 g dry wt cm^{-2} day^{-1}. At the work site during construction mean suspended solids loads ranged from 25 to 60 mg l^{-1}, with peaks up to 1390 mg l^{-1} and sedimentation of up to 0.6 g dry wt cm^{-2} day^{-1}. Downstream of the actual site suspended solids loads increased to give averages of 10–20 mg l^{-1} but sedimentation remained largely unaffected reaching brief peaks of 0.3 g dry wt cm^{-2} day^{-1} during spring run-off peaks. Fish populations, estimated at 52–109 30m^{-1} adjacent to the works site fell to 27 30m^{-1} during the work period. The effects were localized, and 2 km downstream no marked changes were detected. There was no mortality, the losses attributed to migration, and recolonization was rapid with fish populations well established a year later. Benthic invertebrates showed no significant changes in overall numbers or diversity but there were taxonomic changes immediately downstream of the works with mayflies and net-spinning caddis declining and chironomids, mites and cased caddis increasing. Losses were attributed to drift and again recolonization was rapid perhaps from unsilted refuges and by immigration. This example shows the short-term impact suspended solids commonly have, inducing emigration and shifts in the community structure. However, the effects last only for a few months, perhaps due to the lack of any toxicity and spates flushing out the extra sediment. Recolonization by immigration was effective for invertebrates and fish (Table 9.1).

Table 9.1 Short-term effects of inert suspended solids on fish populations. Fish numbers were estimated above, at and below a construction site across a stream that resulted in elevated suspended solids loads before (autumn), during (following spring–autumn) and after (following spring) work. Suspended solids were markedly increased at the site and to a lesser extent downstream. Fish numbers at the site decline, though downstream numbers fluctuate perhaps due to displaced fish moving downstream. By the following spring fish numbers are still lower at the disturbed sites but the suspended solids have declined to previous levels and fish are likely to re-establish. Suspended solids are means (mg l^{-1}), fish number of individuals per 30 metres (After Barton, 1977)

	Before	During	After
Upstream sites			
Suspended solids	3–9	1–10	3–9
Fish population	8–12	6–10	9
Work site			
Suspended solids	<5	25–60	3–8
Fish population	109	55	39
Downstream sites			
Suspended solids	<5	5–20	<5
Fish population	78–81	31–124	40

9.3.2 Metals and mining

Armitage (1980) describes a typical example of heavy metal pollution, associated with old mine workings, on a river benthos. In addition to damage from metals other pollution, for example organic, and natural channel changes show up. The study was based on the River Nent in northern England, which drains an area extensively mined in the past. There has been no active mining since 1943, the study being conducted over 30 years later. Pollution from old mine drainage remained a problem, a typical danger in old mining areas. Samples were taken at points along the channel, sampling riffle habitat to gain as complete a species list of macroinvertebrates as possible and also relative abundance data. Physico-chemical measures included pH, alkalinity, iron, zinc and lead in the water. Preliminary analysis of the invertebrates suggested the sites fell into three categories: low zinc (0.02–0.36 mg l^{-1}), medium zinc (0.77–1.64 mg l^{-1}) and high zinc (2.0–7.6 mg l^{-1}). Natural background zinc levels are between 0.001 and 0.2 mg l^{-1}. All low zinc sites had relatively diverse faunas with stoneflies and mayflies well represented, 14–17 combined species and high abundances of most of them. Total taxa numbered 19–57. Medium zinc sites had fewer total taxa, 13–35, plus low numbers of individuals, mayflies were largely absent. There was a dominance of certain stoneflies and diptera at numbers typically ten times lower than at low zinc sites. Additional stresses were an acidic tributary (pH 3.9, compared to 6.4–7.9 elsewhere in the system) causing local declines and a domestic sewage discharge which appeared to cause a slight increase in taxa, using the extra organic food source. High zinc sites had only 2–15 taxa, with only chironomids and oligochaetes at the site with highest zinc. The animal communities were compared to the physico-chemical measures as a way of characterizing the sites along the system. The communities provided a better discrimination, reflecting the pollution and also natural changes in width and substrate, plus the additional impact of a thick periphyton bloom, a zinc-tolerant alga, *Stigeoclonium tenue,* which all affect the animals.

Examples covering other communities include changes in phytoplankton in a lake exposed to metal effluent; stream periphyton above and below mining operations; impact of mining on litter decomposition, and salmon avoidance reactions, on return migration, to metals from mines. Metals can also accumulate in animals and plants and resistance develops. Good examples include accumulation by bryophytes (Burton and Peterson, 1979) and comparative resistances of algae from polluted rivers (Foster, 1982). Bryophytes have been used as heavy metal pollution indicators for practical monitoring (Benson-Evans and Williams, 1976) as have macrophytes (Abo-Rady, 1980). The chemistry and biology of heavy metals in water has recently been reviewed by Moore and Ramamoorthy (1984; see also Table 9.2).

Table 9.2 Impact of zinc metal pollution from old mine works on stream fauna. Sites along streams in the River Nent system, England, were sampled for invertebrates. Sites differed in zinc pollution. Numbers of invertebrate species present differed between sites reflecting metal pollution, with additional localized problems such as acid inputs degrading some sites that fell into low zinc pollution category. (After Armitage, 1980)

Site type / **Taxa**	Low zinc	Low zinc, acid input	Medium zinc	High zinc
Mayflies	1–10	0	0–3	0
Stoneflies	4–12	0	6–10	0–5
Caddisflies	1–7	0	1–8	0–2
Total taxa	19–57	2–4	13–35	2–15

9.3.3 Thermal

Thermal pollution has been explored in rivers, lakes and at coastal sites, from single species ecology to whole community dynamics. Poff and Matthews (1986) provide a good example of the latter, including analyses of changes to trophic structure. They compared the benthos of three North American streams and the river into which they all flow. One stream was unperturbed, and a second received only limited thermal inputs. An adjacent stream received power station cooling water, not only reaching high temperatures but also fluctuating rapidly. They monitored the conditions during the winter when the thermal regimes in the first two streams ranged from 4 to 8 °C and 4 to 11 °C respectively, that of the polluted stream 7 to 31 °C. They also analysed the river's waters below the confluence of each stream. To look at the benthic communities they used packs of leaves placed into the streams and retrieved at intervals. The benthos washed out of these packs was used as the samples. The standardized quantity and type of detritus allowed straightforward comparisons of the trophic groups present at the different sites. The stressed stream had fewer taxa and fewer numbers of individuals than the other two. Note the loss of both overall taxa present and numbers of individuals, just as with the metal pollution example. A subfamily of the Chironomidae, the Orthocladiinae, dominated the fauna in the unstressed streams, comprising 40–60 per cent of total individuals but accounted for only 5 per cent of the total in the stressed stream. *Physa* snails and the Chironomidae subfamilies Tanypodinae and Chironominae dominated this system. Mayflies and stoneflies were reduced in the polluted stream. Differences also showed up in the trophic functional groups. Collector-gatherers dominated unperturbed sites (65–75 per cent of fauna) but scrapers were most abundant in the polluted stream (39 per cent of fauna). This was attributed to the significantly higher biomass of periphyton in the warmed water and microbial flora on the rapidly decaying leaves in the warmer water. However, trying to pin down a precise cause and effect link is difficult. The fauna may have changed due to alterations in the food supply but it may also be that the grazers happen to be more tolerant of the thermal stress and so survive where the collector-gatherers do not, or most likely a bit of both. Predator numbers were similar in all systems. This study took place during the winter. Pollution, especially thermal, may have a magnified importance at certain times of year, in this case maintaining high temperatures when natural ranges would curtail activity in unperturbed streams. Influence of thermal pollution on duration and extent of ice cover is a special difficulty.

The fauna in the main river below the confluence of the stressed stream was not reduced. The mild thermal elevation, diluted by the river's volume, may benefit the fauna here, with fluctuations dampened and a range of 7–21 °C. The food input may also be increased due to the reduced feeding efficiency by the stressed stream's grazers resulting in a higher export of particulate debris to the river (Fig. 9.1).

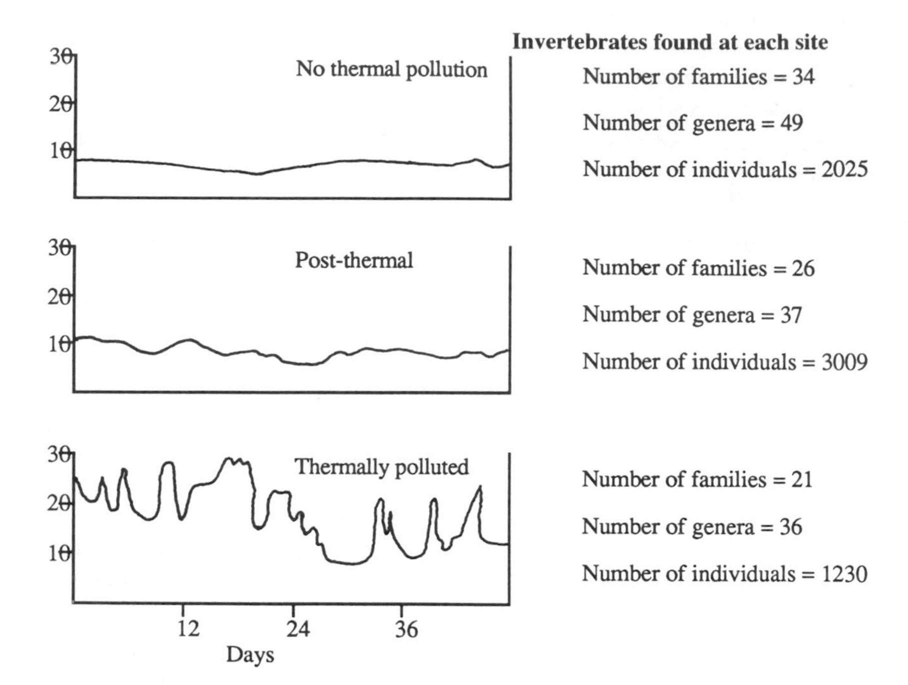

Fig. 9.1 Thermal pollution. The December to February thermal regimes of three adjacent streams are shown: (a) no thermal pollution; (b) a stream that has received no thermal effluent since 1968; (c) a stream receiving thermal pollution. Note that the thermal effluent not only raises the water temperature but also causes rapid fluctuations. Numbers of families, genera and individuals (numbers from five leaf packs) in the three systems are given. (After Poff and Matthews, 1986)

Thermal changes are complex and can be beneficial as well as harmful. The consequences have been extensively reviewed by Krenkel and Parker (1969) and Langford (1983). The latter book includes thermal regime changes caused by impoundment releases.

9.3.4 Pesticides

Pesticide pollution generally arises from the accidental input of chemicals to freshwaters but has also been investigated to look at the effectiveness and side effects of various pesticides for management. Besides immediate toxicological damage subsequent accumulation of residues, including breakdown products, and major alterations to habitat architecture and trophic structure are common problems. The effects of pesticides on stream fauna – and many of the lessons apply equally well to lentic systems – have recently been thoroughly reviewed by Muirhead-Thomson (1987). Newbold (1975) reviewed herbicide impacts on both plants and animals.

A typical example showing toxic effects and longer term consequences is described by Scorgie (1980). Cyanatryn is a herbicide intended for use in freshwaters to treat problem plants. In Scorgie's experiment Cyanatryn was applied to part of a drainage channel in the Fenland of East Anglia. The ditch contained abundant emergent and submerged macrophytes and associated animals. The channel was divided into two by a plastic barrier and one section treated in May, the timing designed to hit the plants at the start of their main growing season. A third section of the ditch further away was also monitored. Slow release pellets were used, made up to give the effective dose of 0.2 $\mu g\ g^{-1}$. Macrophytes, macroinvertebrates, periphyton and water chemistry, including Cyanatryn residues, were monitored before and after treatment, for up to a year, in all three stream sections. Cyanatryn levels in the treated section rose gradually from May to August, reaching a maximum of 0.12 $\mu g\ g^{-1}$. Note how, despite the best calculations, the desired concentration was difficult to achieve, a real problem for careful use of pesticides in aquatic systems. The adjacent screened section was also contaminated by leakage, with levels reaching 0.04 $\mu g\ g^{-1}$. Periphyton, monitored by recording colonization of glass slides was not adversely affected by the treatment with populations in the treated section greatly exceeding those in the untreated, perhaps as a result of macrophyte die-back releasing extra nutrients. Macrophytes were severely affected. Four weeks after treatment submerged swards were visibly sickly and only a few stems remained by eight weeks. After twelve weeks all macrophytes, submerged and emergent, had been lost from the treatment section, resulting in a thick bed of decaying detritus on the bottom. Contamination of the adjacent section resulted in severe losses there. The Cyanatryn concentration seldom exceeded one-tenth of the recommended dose in this stretch but still damage was done. No losses occurred in the third, separate stretch, so all changes could be attributed to the treatment. Regeneration started in the untreated but contaminated section in autumn, and in the treatment section in the following spring. In both, extensive swards reappeared in the following summer but the recolonization included new species and the resulting macrophyte community differed in composition to the pre-treatment one. Careless management of macrophytes, including mechancial removal as well as chemical, has been credited with altering subsequent community patterns, especially by letting invasive, ubiquitous, species gain a foothold and eventually dominate at the expense of sensitive types.

The invertebrate fauna, especially that associated with the dominant pre-treatment water milfoil, *Myriophyllum spicatum* L., changed too. The community was progressively replaced by a benthos assemblage, though total numbers of invertebrates did not decline. The gradual change suggests sudden, toxic effects were not the cause, but that habitat loss and changes in substrate and food source alterations were responsible. The snail *Lymnaea peregra,* which initially made up over 80 per cent of total invertebrate numbers, declined to below 10 per cent. Detritivorous oligochaetes, chironomids, the crustacean *Asellus* and mayflies all increased. These changes occurred in the treatment stretch and in the contaminated, untreated stretch but not in the third, separate section. Analysed using a diversity index, overall diversity actually increased after treatment as the overwhelmingly dominant snail declined and a more even distribution of individuals across different taxa developed.

In this case the communities recovered, though the macrophytes did not return to the previous community pattern. This is an example using a herbicide designed for aquatic work, in careful controlled amounts and still able to cause dramatic changes. The deleterious effects of carelessly used chemicals, including use of those that are quite unsuitable for aquatic habitats can be as deadly as any other toxic pollution and many practical guides to freshwater habitat management urge against use of pesticides, even many of those approved for freshwaters (Lewis and Williams, 1984; see also Fig. 9.2).

9.3.5 Recreation

It is the very attractiveness of freshwaters, their value as an amenity, that places them at risk from recreational damage. The resulting pollution might fall into one of the established categories, for example oil from boat engines or sewage, but physical disturbance, wear and tear and litter are additional threats. They represent a particular threat to sites that are often intact and generally preserved from other pollution such as nature reserves. Liddle and Scorgie (1980) review the effects of recreation on plants and animals. They divided threats into shore- and water-based activities. Water-based dangers derive largely from boating such as wash, turbulence, propeller action, collision and disturbance (all physical damage) and chemical damage particularly from engine oils and sewage. In addition, the threat from pesticide paints on boat hulls and other submerged structures has become increasingly clear. Shore-based damage includes the physical impact of trampling, management practices and disturbance plus chemical pollution from oil, sewage and pesticides. Liddle and Scorgie conclude that the impact on macrophytes was quite well documented, that on animals poorly understood. Since then progress on some aspects has been considerable, notably fishing (Maitland and Turner, 1987). Fishing activities cause general trampling and litter problems but specific dangers stem from discarded line which entangles birds and mammals, lead shot which releases toxic lead into the water and sediments and can be ingested directly by some birds, and

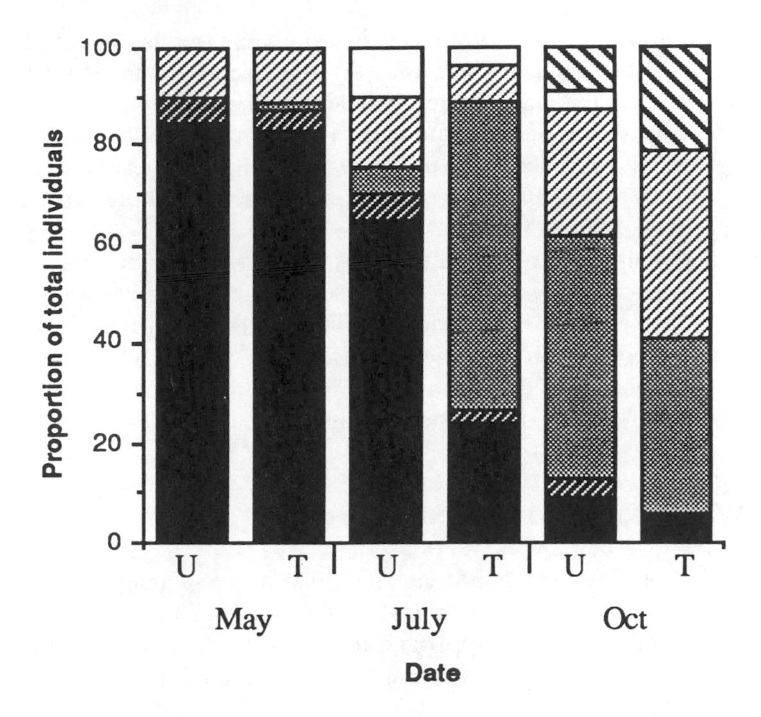

Fig 9.2 Effect of herbicide applicable on macroinvertebrates. The relative proportions of invertebrates in stretches of a ditch treated (T) with the herbicide Cyanatryn or screened off to act as a control (U). Treatment was carried out after May. Note that some contamination of the untreated stretch occurred. Loss of macrophytes resulted in an increase in detritus feeders such as chironomids and oligochaetes. ■, *Lymnaea peregra*; ▨, *Planorbis planorbis* (snail); ▩, Chironomidae; □, *Cloeon* (mayfly); ▧, oligochaete worms; ▧, others. (After Scorgie, 1980)

excessive groundbait as an organic pollutant to the sediment (Cryer and Edwards, 1987).

A detailed study of the effects of boats on macrophytes of canals is presented by Murphy and Eaton (1983). Their study concentrated on canal systems in central England. Macrophytes were sampled to measure relative abundance of species. Boat traffic was quantified using data on traffic through locks, to calculate an index of the number of boat movements, standardized for volume of water, over a year. Macrophyte vegetation fell into four distinct groups, differing in species composition, diversity and abundance. One group was characterized by hardy species such as *Elodea canadensis*, which has many growing points and recovers quickly from disturbance, or *Potamogeton pectinatus,* which is strongly rooted and rather streamlined. This group had the highest index of traffic, with at least 2000, often over 4000, movements a year. The other three groups had more species and lower traffic use, with differences reflecting perhaps natural causes of variation and differences in emergent and submerged plant susceptibilities to damage. Their analysis suggested a distinct threshold at which submerged macrophytes become degraded at 2000–4000 movements per year, with an additional seasonal effect as final community is determined by the growth stage reached by the time traffic attained levels of 300–600 movements per four weeks. The damage was largely attributed to turbidity effects (see Fig. 9.3 and Table 9.3).

Birds and mammals are vulnerable to disturbance, with boats and recent additions such as windsurfing affecting use of waters and breeding success. Batten (1977) describes the impact of sailing on the wildfowl of a London reservoir. He suggests management options, such as discrete refuge areas and distances boats should stay away from birds to reduce the impact and allow multi-purpose use of aquatic habitats.

9.4 Wildlife as pollution indicators

Since pollution causes qualitative and quantitative changes to natural, pristine communities these changes can be used to assess the type and intensity of pollution. This approach has been developed with nearly all forms of wildlife from microbes to mammals. The plants and animals, or lack of them, are used as biological indicators of pollution. Biotic processes such as decomposition rates can also be used. All such techniques are varieties of biological monitoring. Many different schemes have been devised, using all

Table 9.3 Impact of boating activity on canal vegetation. Macrophyte communities were classified using species present and four broad groups were picked out. Average number of species for each group and some typical species positively associated with each group (except group 2 which is very degraded) are shown. The measure of boat traffic is an index of journeys over the entire year at each site. (After Murphy and Eaton, 1983)

	Site group			
	1	**2**	**3**	**4**
Mean number of species	13.6	5.5	11.3	16.9
Mean annual boat traffic	1776.8	5159.2	1817.4	3605.1
Plant species positively associated with group	*Ceratophyllum demersum* *Myriophyllum spicatum* *Potamogeton natans*		*Potamogeton crispus* *P. pectinatus* *Fontinalis antipyretica* *Nitella* spp.	*Sparganium emersum* *Sparganium erectum* *Nuphar lutea*

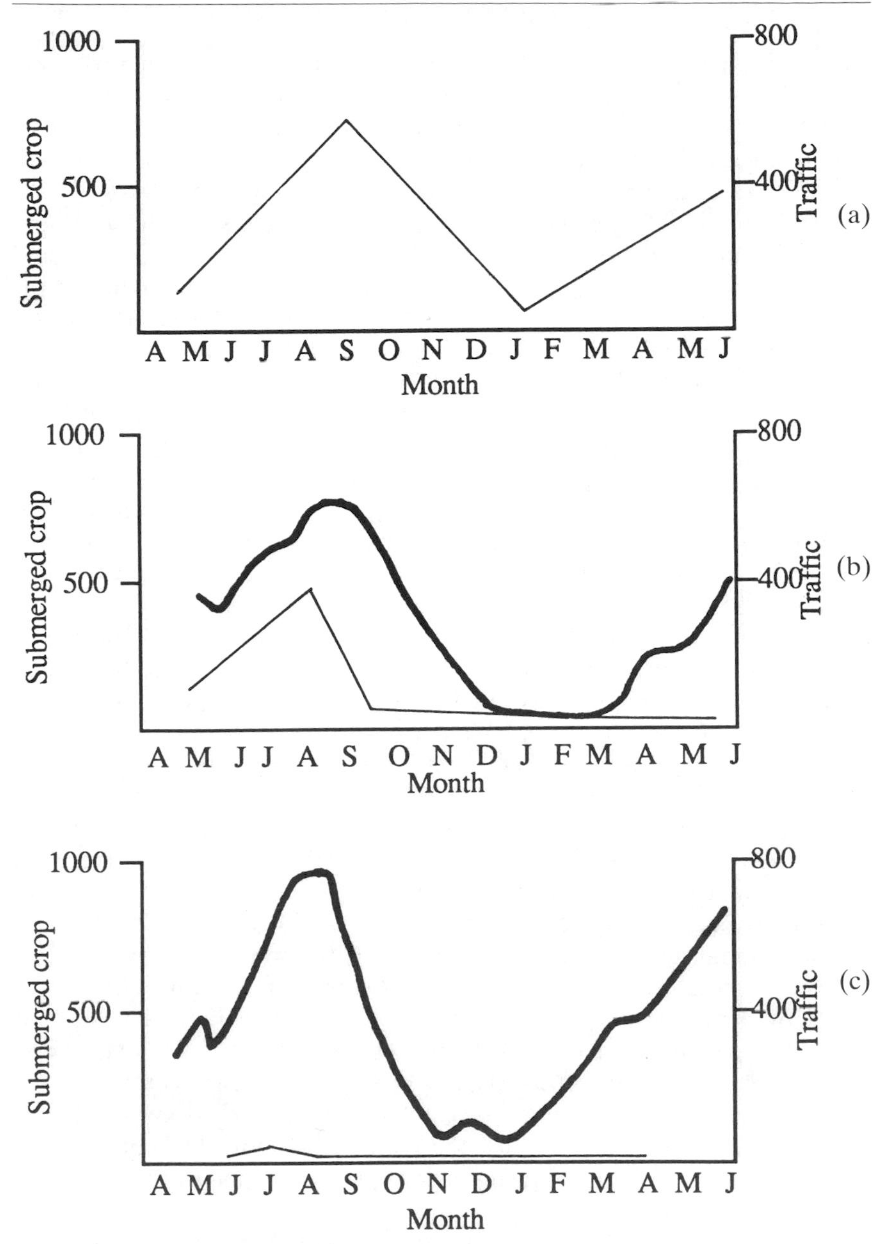

Fig. 9.3 Effect of boat recreation on macrophytes. The graphs show macrophyte standing crop and boat traffic at canal sites with: (a) little or no traffic; (b) traffic reaching the critical threshold for damage; (c) heavy traffic. Plant crop is measured as g m^{-1} fresh weight. Traffic is measured as an index of individual journies over four-week periods. Note how macrophytes decline sharply at critical traffic sites once high traffic levels are reached. (After Murphy and Eaton, 1983)

———— Plant crop. ━━━━ Boat traffic

sorts of wildlife (Hellawell, 1978, 1986; James and Evison, 1979). Biological monitoring has some advantages over actual physico-chemical measures of water quality. The animals or plants respond to all the factors affecting water quality. Physico-chemical measures of everything are impossible, for want of staff, equipment and time, sometimes because we have no way to measure them. The wildlife integrate the combined effects of natural and anthropogenic influences. They do so all year round, continuously, and may still show the effects of past, brief damage events, for example acidic flushes, long after the pollution has gone. The equipment needed to sample and identify the wildlife is often relatively simple and cheap. Biological data provide novel information that the physico-chemical data could not provide, such as the ability of the ecosystem to recover from an incident and incorporate pollutants, such as the use of organic debris or bioaccumulation. However, there are problems. Biological changes do not specify the actual cause, which may be a problem if the pollutant is not known or if natural changes also occur. Sampling must allow for natural patterns, especially standardizing habitat. Riffles are commonly used as standard but not all rivers have typical riffles. Interpretation to wider audiences can be a problem. Either biological information will simply not be understood by untrained administrations and industry or will have to be so simplified as to render it useless. Generally biological and physico-chemical monitoring are used to complement each other. Biological monitoring allows rapid, extensive, general water quality classification. The complex, expensive and limited physico-chemical analyses can then be targeted where problems are apparent.

We have already touched on several biological schemes. Wright *et al.'s*, (1984) classification of invertebrate communities in British rivers (Chapter 5) or Sladecek's (1979) Saprobic Index (Chapter 7) are examples. Here an outline of the systems commonly used in the British water industry is given.

The intent behind most indices is to provide a practical measure of the condition of the water, usually lotic systems. Many indices have been developed but all rely on two basic premises. Firstly, that different taxa vary in their tolerance of pollution. Their presence or absence can be used to gauge the degree of pollution especially if some ranking from sensitive to tolerant taxa is known. Secondly, the numbers of individuals present will also vary with pollution and abundance might also be incorporated into an index of pollution.

In Britain, three indices of pollution, the Trent Biotic Index (TBI), the Chandler score and the Biological Monitoring Working Party (BMWP) score are widely used. The TBI system was one of the first, developed to describe general organic pollution (Woodiwiss, 1964). The index is based on the presence or absence of taxa. A site is sampled to get a tally of the numbers of 'groups' of taxa present. The relevant groups range from each different species present, for some sensitive taxa such as Plecoptera, to grosser levels of classification for some tolerant taxa, such as each family of

Chironomidae present. The total number of groups present, from 0 to 16+ are then cross-referenced against a hierarchy of indicator taxa which range from several species of Plecoptera present, down through a ranking of increasingly tolerant taxa, eventually to not even chironomids or oligochaetes present. The combination of total number of groups and indicator groups gives a score, from 0 (very polluted, no groups not even the most tolerant present) to X (very clean with many groups including several stonefly species). This index is simple. Collection in the field and the level of identification needed make it comparatively easy and fast to use. However, no account is taken of abundances. A major problem is that presence of even one individual indicator organism can raise the final score considerably. It is perfectly possible for a stonefly to turn up in samples from a polluted stretch if it has drifted in from upstream. The animal is not really living in the sample site and may well rapidly expire but will bias the result.

One improvement has been to build in a score for abundance and the Chandler score has been widely used. Taxa are again ranked in a hierarchy from sensitive to tolerant species. A score is given for each, high scores (80–100) for sensitive species down to 1–25 for tolerant taxa. The scores for each taxon also vary across five abundance categories: 1–2, 3–10, 11–50, 50–100, 100+. The Chandler score for a site is obtained by adding up the scores for each taxon present, for the abundance category at which each is found. The scheme has been generally superseded by the BMWP score. This is rather simpler to operate, with a ranking of taxa and a score, from 1–10, given for each one. Samples are sorted and the total score for all the taxa present added up. The precision of taxa tolerances and level to which identification is taken are an improvement on the TBI. In addition to the total BMWP score for a sample the Average Score Per Taxon (ASPT) is often given, dividing the total score by the number of taxa that generated it. A high ASPT reflects the presence of many sensitive species (See Fig. 9.4 and Table 9.4).

The Lincoln Quality Index (LQI) was originally developed in the Lincoln Division of the Anglian Water Authority to meet the requirements of everyday use (Bates *et al.*, 1985). The method uses information provided by both the BMWP score and the ASPT, and combines these in such a way as to give an overall quality rating. Use is made of both the BMWP and ASPT because both measures will respond to variation in water quality and the response to community change will be different in each case. LQI is very easy for the layman to understand, as it uses a simple classification which is directly related to prevailing water quality (Extence *et al.* 1987; Extence and Ferguson, 1989).

All four indices rely on macroinvertebrates. Invertebrates have predominated for biological monitoring over all other wildlife. They are generally easier to sample, at least for qualitative and relative abundance measures. Identification keys are good and their size and equipment needed for identification make this aspect relatively easy. Our knowledge of their

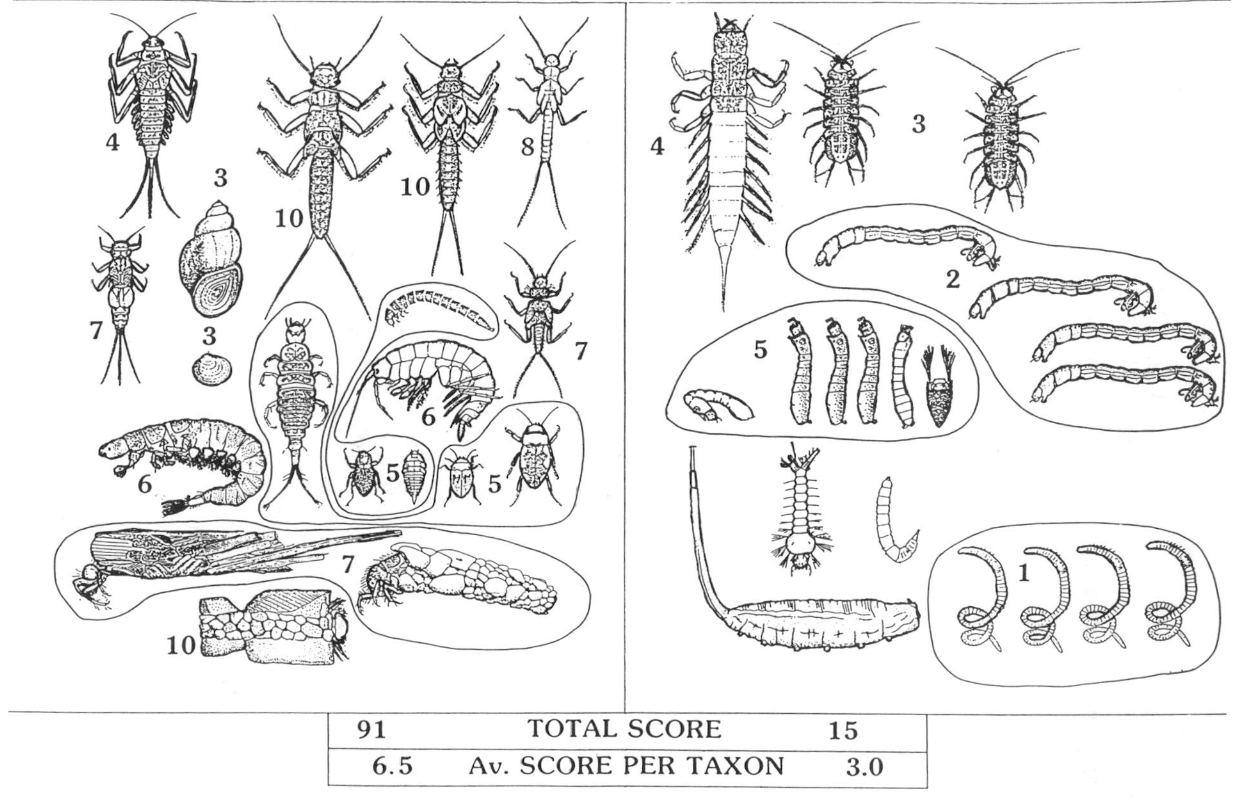

Fig. 9.4 Biological Monitoring Working Party scores (for explanation see text and Table 9.4)

Table 9.4 Examples of taxa scores for Chandler Biotic Index. The Chandler score is worked out by adding up the scores, for those taxa present in a sample at the abundance at which they are found. The stoneflies and mayflies have high score, except for the mayflies of the genera *Baetis* which are more pollution tolerant. Very tolerant taxa such as *Tubifex* worms and air breathing forms such as rat-tailed maggots have low scores, which decline further as their abundance increases. The BMWP score is calculated by adding up the appropriate scores for taxa present, regardless of their abundance in sample. Again stoneflies and mayflies have high scores, except *Baetis*. The total BMWP score is commonly divided by the number of taxa contributing to the total to give the Average Score per Taxon (ASPT)

	Chandler score Abundance in sample					**BMWP** taxa score (scores are for members of families)
Selected taxa	1–2	3–10	11–50	51–100	100+	
Each species of Perlidae, Perlodidae, Isoperlidae, Chloroperlidae and Taenoptergidae	90	94	98	99	100	10
Each species of Ephemeroptera, excluding *Baetis*	79	84	90	94	97	10 (Caenidae 7)
Each species of *Baetis*	44	46	48	50	52	4
Each species of *Tubifex*	22	18	13	12	9	1 (whole class)
Each species of air breathing life	19	15	9	5	1	Not specified
No animal life			0			

tolerances and responses to pollution is extensive. They are also a diverse group, with tolerances varying between species allowing discrete differences to show up and containing both sensitive and tolerant indicators. Problems include insensitivity to some pollutants and seasonal changes, notably the summer hatch of many taxa, especially sensitive groups such as stoneflies, which are then absent though not because of pollution. Macrophytes tend to be rather widely tolerant of pollution, communities lacking the finely tuned responses of invertebrates. Development varies greatly across season. However, considerable effort has been put in to develop macrophytes as a tool to assess general condition and degradation of rivers, including engineering and recreational impacts in Britain (Haslam and Wolseley, 1981). Algae are as diverse and sensitive as invertebrates, but sampling and identification are difficult. Fish are difficult to capture and their mobility results in avoidance of polluted stretches so they are either present or absent with little gradation in between. However, their species composition may change, with those tolerant of low dissolved oxygen levels and high water temperature (e.g. cyprinids) remaining after the more sensitive species (e.g. salmonids) have disappeared.

Recent developments of multivariate community analyses, integrating whole catchment processes, have provided the basis for further progress. Analysis of the whole communities incorporating natural processes and pollution damage is now a possibility currently being developed in the British water industry. Biological monitoring will remain an essential part of water quality assessment and management.

10 Effects of land use

The chief forms of land use are agriculture and forestry and there are many agricultural and forestry practices which may have either a direct or an indirect effect on aquatic habitats.

10.1 Agriculture

10.1.1 Land drainage

One of the main agricultural practices which has an adverse effect on both streams and ponds is land drainage. Land drainage, often involving river channel works or channelization, affects many upland streams which are the habitat of young salmon and trout. In these areas land drainage is undertaken to improve farmland by land reclamation and to assist in flood control. For these reasons many streams being modified or 'improved' are straightened to reduce meanders and are dredged to deepen the stream channel so as to accommodate the predicted increased flow. As a result there is a reduction in the size and diversity of stream habitat. Instream cover in the form of boulders and large stones is removed, the pattern of alternating riffles and pools is drastically altered and the substrate is modified. Furthermore, riparian and bank habitat is changed with the loss of vegetation. As a consequence, the initial effect on the stream is to reduce the size of the fish populations and the size and diversity of invertebrate communities. Frequently there is subsequent erosion due to the shape and nature of the realigned banks and increased water turbidity occurs at high flows, resulting in sedimentation of the stream with a consequent smothering of the substrate and a loss of invertebrates, fish spawning areas and young fish habitat. Many of the larger pools may be infilled with gravel washed downstream due to the unstable conditions arising from dredging. This results in the loss of holding pools for larger fish.

Stewart (1963) refers to the drainage of the fells in the upper reaches of the Ribble in Lancashire. In 1947 small open channels 30 cm wide and 23 cm

Plate 22 A Cuthbertson drain showing effects of severe erosion to a depth of over 2 m only two years after ploughing. The eroded material is carried into a major salmon spawning tributary of the River Oykel in east Sutherland (see Chapter 10). (Photo: Derek Mills)

deep were cut in the hills in the form of a herring bone. By 1963 the main channels had eroded into large drainage channels varying in width from 2 to 3.5 m with depths greater than 1 m. Stewart found that the amount of sediment flowing down the River Lune in Lancashire during a spate amounted to 1½ tonnes for every million litres of water. Wolf (1961) has also shown in the case of the Kavlinge river system in Sweden that, through ditching, canalizing and cultivation of bad as well as good farmland over 150 years, the greater part of the surface has disappeared. The intensive cultivation of the area has caused every possible bit of land to be drained, ditched and ploughed. He estimated the amount of topsoil in the catchment of the Kavlinge river system carried away by flood water. He found that 1 m of water contained about 50 g of humus material and about 100 g of minerals. This meant that a stream carrying 60 m of water per second also carried with it 3 kg of humus material and 6 kg of mineral matter. So in 24 hours about 780 000 kg of soil would be carried downstream.

Spillett *et al.* (1985) describe some of the amieliorative methods to reinstate fisheries following land drainage operations in River Thames tributaries. These included installing stone gabion weirs, groynes and deflectors.

10.1.2 Channelization

As Brookes (1988) stresses, stream channelization is an extreme physical disturbance that upsets the whole stream ecosystem. If channelization disturbs a first-order stream, allochthonous organic particle input is reduced or eliminated with removal of bank cover. The channel is open to direct solar input, and the essentials for autochthonous communities develop.

In the early 1980s a number of farms in the south of Scotland carried out extensive land reclamation involving straightening and deepening of stream channels. On one tributary of the River Tweed 162 ha of land were reclaimed. A survey of this modified section of stream, where salmon used to spawn, revealed a total absence of young salmon. A later survey in 1987 showed that salmon were beginning to recolonize this area. A similar situation occurred on the River Camowen in Co. Tyrone, Northern Ireland, where dredging operations initially reduced densities of young salmon but where there was a subsequent progressive recovery. The effect of arterial drainage on the Trimblestown River on the Boyne catchment was a change in fish species from predominantly salmonids to small riverine coarse fish species (McCarthy, 1983).

Channelization has been shown to severely reduce the standing crop and diversity of fish populations of streams in several regions of the United States. A study of 23 channelized and 36 natural streams in North Carolina revealed that channelization reduced the number of game fish over 15 cm in length by 90 per cent and reduced the weight by 80 per cent (Bayless and Smith, 1967). Only limited recovery was observed 40 years after channelization.

Dredging of streams can also have other drastic effects. In Finland, for example, dredging for timber floating has altered the natural structure of the river bed of many northern rivers by removing the coarser bottom material and by changing the water flow in the rapids.

A number of recommendations have been made for ameliorating the adverse effects of channelization (McCarthy, 1985):

1. Raking of the gravel to remove the accumulated sediment.
2. Installation of instream structures such as random rock clusters, jetties, low dams, weirs and current deflectors.
3. Stabilization of the riparian zone through removal of spoil banks and revegetation schemes.
4. Replacement of gravel in riffle habitats.

It should not be forgotten that ponds, as well as streams, are affected by land drainage operations and many farm ponds have dried up as a result of these activities. In many instances their disappearance, with their varied plant and animal communities, is a greater loss than those existing in the countless miles of remaining stream.

10.1.3 Sheep grazing and muir burn

The main effect of sheep grazing and muir burn is soil erosion which influences the aquatic environment through consequent sedimentation. In Scotland and Iceland sheep cause a great deal of erosion by destroying the ground vegetation by grazing. They also contribute to the instability of hill ground and river banks by creating innumerable tracks and narrow paths and by using small knolls and irregularities in the ground and river banks for protection from the weather; these places are gradually worn down in the form of 'scrapes' until a shallow soil profile is exposed which later increases in width and depth. Fairbairn (1967) and McVean and Lockie (1969) mention just how dangerous this erosion can be by drawing attention to the south-east Scotland floods of 1948 and floods in the White Esk, Ettrick, Dulnain and Lochaber areas in 1953.

Muir burn may contribute to erosion and to landslides in certain types of terrain. Muir burn is an age-long practice of rotational firing of heath land to promote new growth of ling heather (*Calluna vulgaris*) which provides part of the diet of sheep and grouse. The firing is usually carried out in the spring as a part of basic moorland management. Fairbairn (1967) points out that repeated burning at close intervals may result in the disappearance of herbaceous species, which in turn has an adverse effect on the soil, destroying the organic horizon, whereupon erosion is inevitable and becomes accelerated. Encouragement of heather moorland has also been associated with changes in acidification. To reduce acidification, tests have been made with various remedial treatments, one of which involves burning of heather moorland to release alkaline material to the soil.

10.1.4 Farm wastes

Some of the major farm wastes come in the form of slurry from dairy farms and piggeries and silage liquor, all of which are grossly polluting being high in oxygen-demanding properties. This source of pollution appears to be increasing. In 1986 dairy farms in England and Wales made up the bulk of the offenders in the United Kingdom with 2400 incidents out of a total of 3427. In south-west England Merry (1985) refers to increases in fish kills caused by farm wastes. In one incident in 1982, 32 km of the River Axe were affected. On the River Torridge salmon spawning areas have been rendered useless and even cattle will not drink from the affected parts of the stream. It is considered that Northern Ireland is proportionately the worst area in Britain for silage pollution offences (Kennedy, 1987), while the Republic of Ireland has insuperable problems in the Lough Sheelin area over the disposal of piggery wastes.

These pollution incidents usually result from the inadequate storage of farm slurry in tanks and lagoons. Often the pollution occurs after heavy rain or thunderstorms when these ponds or lagoons burst their banks and storage tanks overflow Silage. Liquor is a very corrosive liquid and will often leak from clamps and towers into streams.

One way of overcoming the farm waste problem is greater publicity and awareness among the farming community and more available advice. To this end, the South-West Water Authority in England has produced explanatory leaflets for the farmers and visits are arranged to explain the situation. This is no mean task as there are more than 15 000 farms in the water authority's area. So a pilot scheme was started to deal with small parts of the Rivers Taw, Torridge, Axe and other catchments with known farm drainage problems. The scheme was supported fully by agricultural interests such as the Country Landowners' Association, the Agricultural Development Advisory Service (ADAS), the National Farmers' Union (NFU) and the Farming, Forestry and Wildlife Advisory Group (FFWAG).

Howells and Merriman (1986) put forward the following remedies which, if these objectives were achieved, would result in a better appreciation by the farmers for the need for proper handling and storage of wastes and better installation of suitable systems at the outset of any improvement scheme: (a) full consultation at the planning stage; (b) farmers should be fully informed of the polluting nature of farm wastes, including chemicals, and the best ways to contain, dispose of, or treat these wastes to minimize pollution; (c) measures to store and dispose of waste materials on farms should be eligible for grant aid.

10.2 Forestry

The upland areas of Great Britain and many other European countries are being extensively afforested. According to Gill (1989) the proportion of land under forest in Britain has doubled since 1919 from 5 to 10 per cent to a current area of 2.3 million hectares. The expansion of forest in the last five years has been at an annual average rate of 24 000 hectares. It is the stated intention of the government that this expansion should continue with an annual target rate of 33 000 hectares. It is in these upland areas that many salmon and trout rivers begin, arising from an extensive network of small, fast-flowing tributaries which serve as the spawning and nursery areas. These upland areas are invaluable to the river system, and the degradation of their abundant, clean, cool, well-aerated and silt-free waters would lead to the loss of much of the area upon which the Atlantic salmon and trout depend for the survival of their young. It is not always realized that even the smallest of these streams, some little more than one metre wide, can be important. It is therefore essential to realize how delicate the natural balance is when considering various forest management practices which could affect the aquatic environment.

10.2.1 Road construction

Often to gain access to the land for ploughing and planting roads require to be built. These are unmetalled and usually no more than tracks of bare earth and stone. Consequently when it rains there is a run-off of suspended material which ends up in the nearest watercourse and sedimentation from this source can be insidious and potentially serious. To overcome these effects riparian vegetation should be protected and sediment and debris prevented from reaching the streams. Drains should run alongside the road and have no direct discharge into natural watercourses. Streams in the route of the road should be culverted and culverts should be of sufficient size for the run-off from the drainage area.

10.2.2 Ploughing and draining

Sedimentation, resulting in silting of fish spawning and nursery areas and infilling of holding pools, can occur from erosion of plough furrows and drainage channels on steep hillsides. Alignment of ploughing along the contour was once favoured and is still occasionally practised. It is now felt that this is inappropriate in wetland situations, since each furrow tends to

Plate 23 Forestry drainage operations leading to loss of bank cover, erosion, sedimentation and movement of gravel (see Chapter 10). (Photo: R. McMichael)

'pond' surface run-off and gives rise to lines of stagnant conditions. Alignments are now made to run roughly up and down the slope to improve run-off and provide pathways for downhill seepage when the furrows become filled with plant litter. The risk of severe erosion has been shown to be small, provided cut-off drains are used and individual furrows are not used to carry water drainage from large areas lying above.

The alignment of the drain should be designed to achieve the maximum interception effect with the minimum drain length. In order to catch silt where drains are leading to streams or small watercourses they should either be tapered in depth and stopped 15 to 20 m short in order that the water discharging from these drains has to filter through the ground vegetation or, in areas of high rainfall, a sump should be taken out just before the drain opens into the watercourse. From time to time it may be necessary to clean out drains within the forest if waterlogging is evident. This operation should be very carefully timed so that little silt enters the water courses to affect the stream bed and bank habitats. For this reason, it is important not to clean out drains during the period mid-October to mid-May, if possible, as during this time the eggs of salmon and trout, and latterly the newly hatched fish, known as alevins, are in the gravel and would be suffocated by the silt.

10.2.3 Planting

The layout of planting and the choice and location of tree species have a great influence on adjoining aquatic environments. Changes in the flow regime of streams, as well as being affected by drainage, are further accentuated by higher water use by a forest through transpiration and it has been shown that trees, under certain conditions, intercept substantially more water than open moorlands in areas of high rainfall.

Planting of conifers close to the edges of streams seriously affects the stream's productivity, altering both the autochthonous productivity and allochthonous inputs. This leads to a paucity of insect and fish life in the streams which flow through mature forest with a dense tree canopy, which prevents the growth of ground vegetation. Bank vegetation is important in providing hiding places for vertebrates and invertebrates and also helps to maintain deep pools by preventing lateral erosion, which results in a wider but shallower stream channel. Mills (1969) recorded the densities and standing crops of brown trout in a stream in southern Scotland which flowed through immature and mature coniferous plantations (see Table 10.1). It was noticed that in those sections of stream which flowed through immature forest with dense vegetation overhanging the bank the standing crop of trout was much higher than in those parts of the stream which ran through mature forest where the absence of streamside vegetation had led to considerable erosion of the banks, resulting in less cover and the loss of an important part of the input of allochthonous material. Smith (1980) also found in a study

Table 10.1 Standing crops of brown trout related to streamside vegetation in a forest stream. Figures in brackets denote annual production: *, all trout were removed the previous autumn; **, 200 metres downstream of section 4. (*Source*: Mills, 1969)

Section	**Streamside vegetation**	Area (m^2)	Total length (m) (sections 0–4)	**Standing crop (kg ha^{-1})** Oct. '66	June '67	Oct. '67
0	10–20-year old Douglas fir and Norway and Sitka spruce	101		—	—	238 (121)
Fire Dam						
1	Abundant undergrowth over-hanging stream.	209		192 (90)	—	107* (90)
2		90	1118	86 (45)	—	93* (45)
3	49-year-old Douglas fir and Norway and Sitka spruce	552		56 (30)	—	27* (30)
Trap						
4	No undergrowth	186		51 (30)	56 (30)	—
5**	Grazed grassland and *Juncus* spp.	124		—	100 (45)	—

Plate 24 A forest stream running through mature forest (49 years old) of Douglas fir and Norway and Sitka spruce with a closed tree canopy. Absence of riparian vegetation has led to considerable erosion of the banks with consequent loss of cover (see Chapter 10). (Photo: A. Grandison)

of a partially afforested stream in southern Scotland that invertebrates and fish were considerably less abundant where the stream was densely shaded by coniferous trees. This is revealed in Fig. 10.1 which shows the total number of invertebrates and taxa collected in the spring, summer and autumn at each of the main sampling stations on this stream, which covers sections flowing through moorland, coniferous forest, meadow and deciduous spinney.

A dense tree canopy also reduces the extremes in water temperature, so that in winter average water temperatures are higher than in an open area of stream but in summer they are much lower. These differences in water temperature affect the growth of fish and may influence the length of the egg and larval or nymphal stages of some aquatic insect groups.

Adverse effects of dense streamside tree canopy can be reversed or minimized by the creation of reserve strips, or unafforested buffer zones, between the plantation and the banks of the watercourse. The protective strips should ensure that at least 50 per cent of the stream is open to sunlight with the remainder under intermittent shade from shrubs and light-foliaged

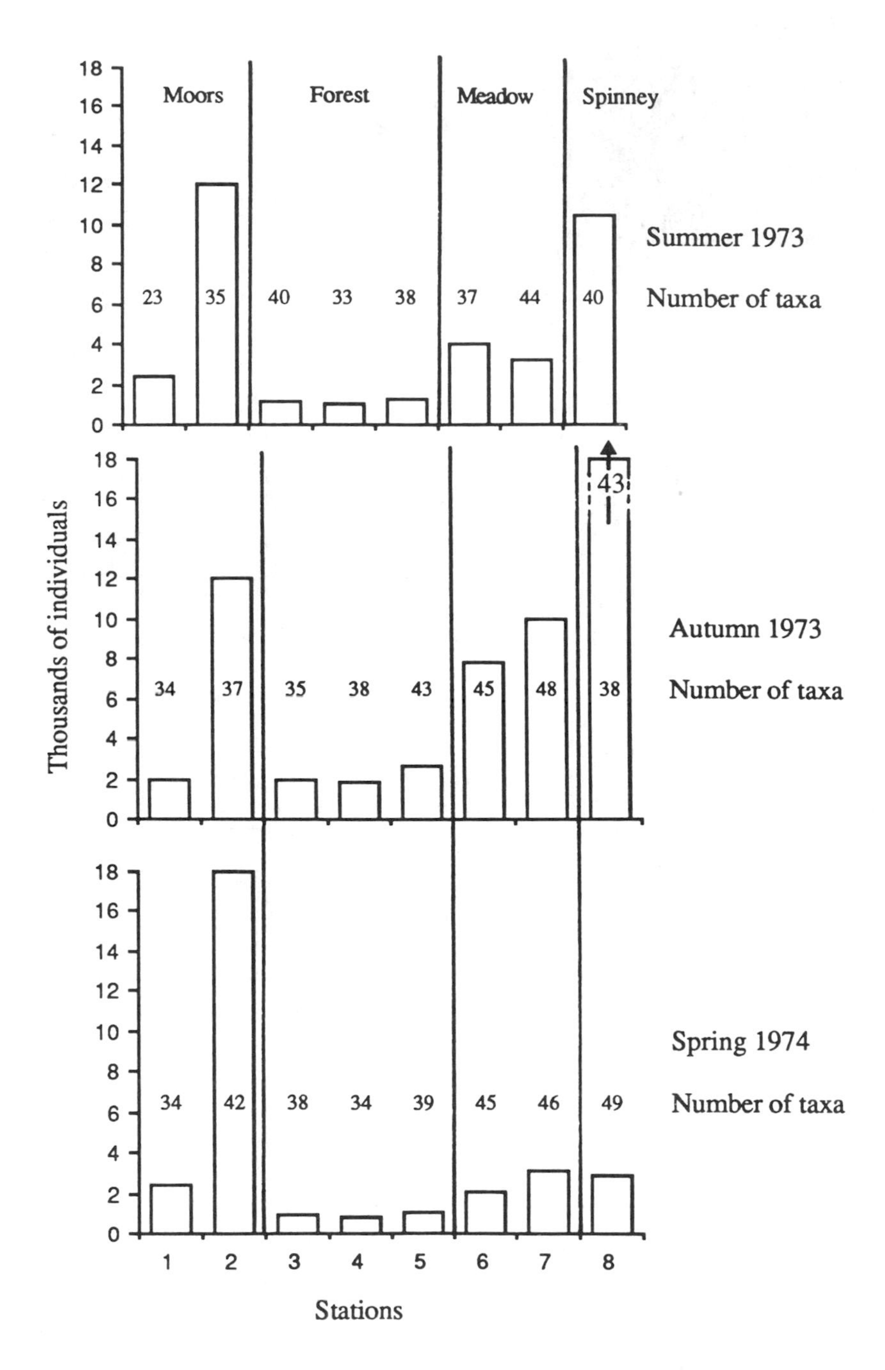

Fig. 10.1 The total number of invertebrates and taxa collected at each of the main sampling stations. (*Source*: Smith, 1980)

trees such as ash, aspen, birch, hazel, rowan and willow. The leaves from these trees will provide a source of food for fish in the form of a 'fall' of terrestrial insects, dead plant material for detritus feeders in the invertebrate community and a source of nutrients for the stream.

10.2.4 Application of fertilizers

Leaching of fertilizers into streams has been well recorded (Harriman, 1978). In some cases this leaching has had little effect on water quality while in other instances it has been suggested that it could increase stream productivity and so be beneficial. However, leaching into streams running into lakes and reservoirs could produce undesirable conditions of eutrophication.

10.2.5 Spraying

The effects on fish and stream insect populations of improperly controlled forest spraying can be very serious. The spraying of spruce budworm with DDT in New Brunswick did considerable damage to the stocks of young salmon in the Miramichi River watershed. After the banning of the use of DDT in Canadian forests in 1968 Fenitrothion was used instead. Although this had no direct effects on fish it did reduce some of the aquatic invertebrate populations, the arthropods in particular. This reduced the food supply for young salmon whose growth was consequently reduced. It is common knowledge that spraying over watercourses should be avoided if at all possible, and that the correct spray dosages should be used to minimize the damage to fish and wildlife. When agricultural chemicals are used at recommended rates and with due care the risk to aquatic animals is relatively small but not entirely eliminated.

10.2.6 Thinning, felling and extraction

The main effects of these operations on the stream environment can be changes in stream flow and water temperature, sedimentation, loss of nutrients, damage to spawning grounds, blockage of streams by felled timber and branches, prevention of the movement of migratory fish and bacterial decomposition of any bark and wood debris smothering the stream bed.

Trees should be felled away from streams. If tree tops and branches do enter a stream they should be removed as soon as possible. When logs are moved care should be taken not to break the soil surface of the forest floor unduly, resulting in stream sedimentation.

Tree removal also leads to an increase in water temperature due to lack of shade. A study in the Smoky Mountains of Tennessee showed that trout moved upstream to cooler water following logging of the watershed (Greene, 1950), and in coastal streams in Oregon Hall and Lantz (1969) found not only a substantial change in water temperature but also in the dissolved oxygen content of the water following logging.

During logging operations logs may be stranded and direct the water flow from gravel bars. This may result in the drying out of deposited spawn, or diversion of normal water flows from potential spawning areas. Log jams and logging debris often cause obstructions and, where they hinder the migration of fish, are undesirable. However, unless jams are impassable or at least a hindrance to fish passage, they may be a stream asset as the deep holes usually associated with jams afford places of refuge for salmonid fish prior to spawning.

The majority of the above-mentioned forestry operations have complex effects upon aquatic environments, and this is particularly marked for clear-felling. This operation alone produces a decrease in transpiration and interception resulting in a change in the flow regime; increased solar radiation leading to higher water temperatures: increased primary production, and nitrification resulting in increased primary production and eutrophication.

A pictorial description of the various procedures to adopt to safeguard the aquatic environment in an afforested area is depicted in Fig. 10.2, which appears in *Forests and Water: Guidelines* produced by the Forestry Commission (1989).

A Establish broadleaved trees near watercourses.

B Maintain about half of stream surface in full sunlight, the rest in dapple shade.

C Stop plough furrows well short of watercourses.

D Use water for conservation.

E Do not plough unnecessarily; consider scarifying or mounding.

F Maintain protective unplanted vegetation strips not less than 5 m wide on each bank.

G Keep branches and tops out of stream.

H Stack timber on dry ground away from watercourses.

I Design streamside edges in harmony with the landscape.

Fig. 10.2 Procedures to adopt to safeguard the aquatic environment in an afforested area.

11 Fish Farms

Fish farming has been carried out for hundreds of years and first started in China with the rearing of species such as carp (*Cyprinus carpio* L.). The rearing of fish, chiefly cyprinids and cichlids, in captivity in ponds, swamps and paddy fields, continued to develop in South-East Asia, Egypt, parts of Africa and Eastern Europe. Since the end of the last century fish farming started to increase in Western Europe and North America with the development of salmon and trout farming, particularly the latter, using earth and concrete ponds, tanks and raceways which are provided with a continuous supply of good quality water. These cold water species of fish are more demanding in the culture conditions they require compared to fish such as the carps, tilapias and catfishes, which can tolerate a wide range of water temperatures and low levels of dissolved oxygen.

11.1 Fish ponds as artificial communities

The basic principle of fish culture (Fig. 11.1) is not unlike a natural lentic system which can be manipulated in several ways by altering the inputs. It involves the introduction of fish fry to an enclosed area of water and growing them on to a marketable size, sometimes hastening their growth rate by supplementary feeding. This addition of food to the pond is necessary if the density of fish is higher than the pond can support naturally. Sometimes it is the practice to increase the productivity of the pond by adding extra nutrients in the form of cow or pig manure, dry poultry waste or artificial agricultural fertilizers. This is practised particularly in Taiwan and Israel where fish farm units are integrated with general agricultural activities such as animal husbandry (Hepher and Pruginin, 1981). Pond fertilization can also be achieved by farming ducks on the ponds, as is done in Malaysia, Taiwan, Eastern Europe and The Netherlands. The duck droppings are an excellent source of fertilizer. Human sewage is also a very good source of nutrients and carp can be used as efficient sewage convertors. In Munich, in Bavaria, the settled and partially treated sewage of the city is diluted and led

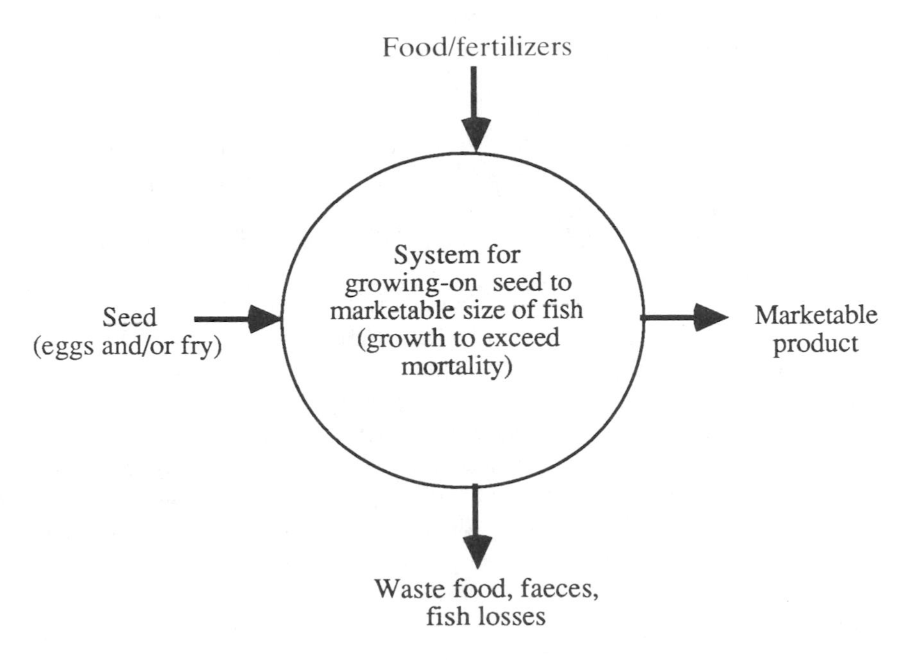

Fig. 11.1 Basic principle of fish culture.

Plate 25 Collecting the seed. Dubisch or Hofer ponds at Wielenbach Fish Culture Station, Bavaria. These are spawning ponds. The mature carp are put in the ponds until they have laid their eggs. The resulting fry are left in the ponds for some days and then transferred to larger rearing ponds. (Photo: Derek Mills)

through a 7 km series of 4 to 5 ha ponds, each containing about 5000 two-year-old carp which are fattened on the abundant invertebrate fauna present. The ponds are in fact being eutrophicated, with the high productivity being put to good use.

Many ponds are capable of being drained using a special sluice known as a 'monk' with which it is possible to adjust the water level in the pond. The fish collect in a specially designed basin behind the 'monk' and can be netted out. The pond can then be left to dry out, after which fertilizer can be added and the pond then refilled.

A further way of increasing the yield of fish from the pond is to rear a variety of fish species with different feeding habits and occupying different trophic niches so that they can utilize more fully the whole plant and animal community living in the pond. For example, common carp, tilapia (*Oreochromis* spp.), silver carp (*Hypothalmichthys molitrix* L.), grass carp (*Ctenopharyngodon idella*) and, to a certain extent, mullet (*Mugil* spp.) and bighead carp (*Aristichthys nobilis* L.) all differ in their feeding habits and their concurrent culture in a pond increases total yield. This practice is known as polyculture while the rearing of only one species is referred to as monoculture.

There are, of course, as in natural aquatic systems, other factors limiting the holding capacity of a fish farm. One of these is the availability of dissolved oxygen. This limitation can be overcome by supplying extra oxygen to the water by means of aeration. Water temperature is also important and various fish species have optimum water temperature requirements. Rainbow trout, for example, have an optimum temperature range for food conversion of 14 to 16 °C.

A nice example of a fish pond system in Malaysia is given in Fig. 11.2. In this system a pig farm is sited alongside ponds stocked with carp. The carp feed naturally in the pond but their diet is supplemented with waste food from the piggery. Extra nutrients also enter the ponds from the piggery and are utilized by the pond ecosystem and help the growth of a commercially important water plant known as water kang-kong which is both eaten by the carp and harvested by the farmer. The farmer therefore has three crops – pigs, carp and water kang-kong.

Another example of a multiple-use system involving fish is one adopted in China. This is the growing of mulberry trees and the raising of silk moths: the silk moths feed on the mulberry leaves alongside the carp ponds. The mulberries fall into the water and are eaten by the carp. When the silkworms (the caterpillars of the silk moth) come to pupate the silken cocoons are removed and the silk is spun from them. The unwanted pupae are then used as a supplementary feed for the carp (Hickling, 1971).

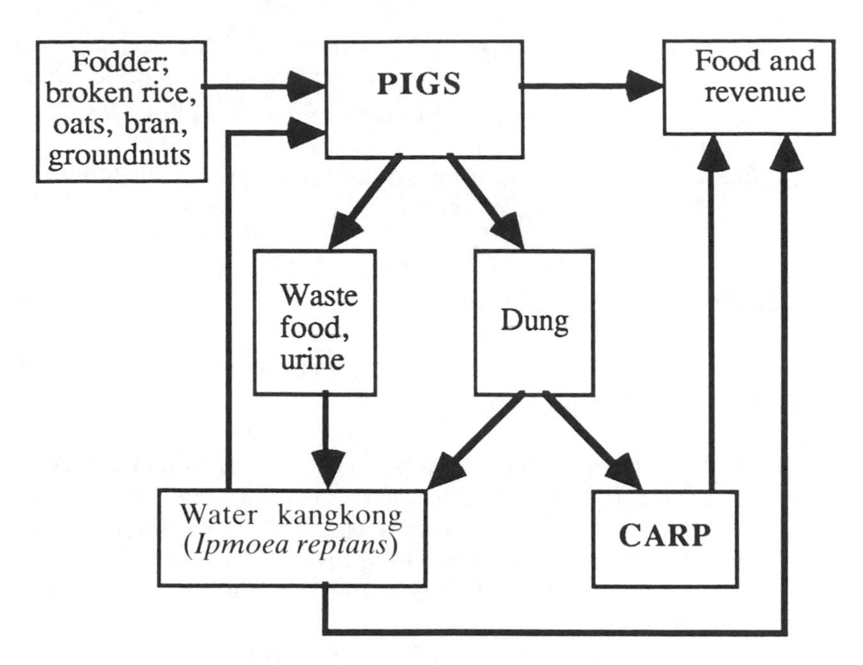

Fig. 11.2 Pig–carp culture.

11.2 Integrated lake farming

An interesting recent development in China has been integrated lake farming for fish and environmental management in large shallow lakes. The two major components of integrated lake farming are aquatic plants and fish. The concept of integrated lake farming is similar to integrated land farming (e.g. raising cattle) in which pasture is raised to feed the cattle and the manure from the animals is, in turn, used to fertilize the pasture. Aquatic plants are selected on the characteristics of high efficiency in using solar energy and serving as both food for humans and feeds for fish. The plants used are free-floating emergent macrophytes as they are better than phytoplankton at fixing solar energy. The plants are grown in what are referred to as hydro-agriculture fields. Seedlings are transplanted from nurseries to growth nets suspended in the lake. The main plants used are the water spinach (*Ipomaea aquatica* forsk) and trapa (*Trapa natans* L.) both of which are used as human food. The plants are harvested and a proportion are fed to the fish held in nearby enclosures in the lake. The grass carp is the most important fish used in this lake farming (Chang, 1989).

This integrated lake farming can be a useful method for reducing the effects of eutrophication. Because shallow lakes are easily mixed by wind the sediment is frequently resuspended when wind and current are strong. This causes the nutrients in the sediment to be released into the water column, producing frequent algal blooms. The extensive planting of macrophytes for lake farming stabilizes the sediment and reduces nutrient resuspension. The result is a reduction in the frequency of blue–green algal blooms in areas where extensive aquatic vegetation is planted.

The whole subject of fish culture is admirably covered in such texts as those by Huet (1986) and Bardach *et al.* (1972).

11.3 Environmental problems

Not all the young fish added to the system described above (Fig. 11.1) reach a marketable size and the mortality of the young fish contributes to the waste emanating from the pond. If there is no outfall from the pond the waste, in the form of dead fish, excretory products, faeces and unconsumed food, is recycled by the breakdown by bacteria and fungi. However, if there is a pond outlet, the waste products are released to the outside natural environment and can cause environmental problems.

These problems occur chiefly in the culture of salmonid fish. This is because they require large volumes of water which is fed from a neighbouring river through either earth or concrete ponds, raceways or fibre glass tanks. The outlets from these ponds and tanks also discharge relatively large amounts of water heavily contaminated with waste food and fish faeces. Suspended organic material from these ponds and tanks gives rise to a number of problems. As suspended solids, they smother the stream bed, alter the invertebrate community and may cause sewage fungus to develop. They frequently discolour the receiving water making it useless for any other purposes. As reservoirs of plant nutrients they encourage eutrophication, yielding one-third of their phosphorus content in two days. Solbé (1982) recorded a number of changes which occurred to water during its passage through a fish farm. On average, the dissolved oxygen concentration fell by 1.6 mg l^{-1} and there were increases in biochemical oxygen demand of 1.5 mg l^{-1}. Although these are relatively small changes the mass flows from the farms he sampled were large enough to change the concentration in the receiving waters. A decrease in dissolved oxygen of 0.3 mg l^{-1} occurred in the river and there were increases in BOD of 0.7 mg l^{-1}. Suspended solid loads also increased significantly and the net mass outflow of suspended solids per tonne of fish produced has ranged from 1.35 t in 1980 to 0.55 t in 1987 (Solbé, 1987).

Chemicals and medicaments used in fish disease prevention are also released from the ponds and may affect the stream fauna if present in sufficiently high concentrations. These include: (a) anti-foulants on nets and cages which, until its ban in the mid-1980s, was usually tributyltin (TBT); (b) disinfectants, such as formalin and malachite green, used to combat parasites and diseases; (c) food additives, including vitamins, minerals, pigments and hormones; and (d) antibiotics to combat disease.

Plate 26 Another method of collecting seed. An artificial spawning nest or kakaban, held here by the fish farmer, is placed in the pond as a spawning substrate for the fish. When the kakaban is covered with eggs, in this case those of the zander or pike-perch (*Stizostedion lucioperca* L.), it is taken into a hatchery where it may be suspended over a tank under a fine shower of water. The fry drop into the tank on hatching. They are then transferred to small rearing ponds or tanks. The location is Rosenheim in Bavaria. (Photo: Derek Mills)

Solids can be removed from fish farm effluents by settlement lagoons where the risk of eutrophication in the receiving water is not an issue, or by swirl concentrators or triangular filters (Solbé, 1987).

In recent years some countries, such as the United Kingdom and the United States, have started to rear trout and young salmon and, in the case of the United States, channel catfish, in cages moored in lakes, lochs and reservoirs. This overcomes the problem of water supply but inevitably raises other problems. Most of these emanate from the input of solid waste as (a) suspended solids deposited below the cages and (b) dissolved solids, particularly phosphorus, which may affect the nutrient status of the water body (Fig. 11.3). In the United Kingdom a large number of upland lakes are used to store water for domestic and industrial supply. These lakes are often oligotrophic and must be kept in that state as oligotrophic water costs little to treat. The problems have been fairly extensively studied in the cage culture of rainbow trout in fresh waters (Beveridge, 1984; Phillips, 1984, 1985; Merican and Phillips, 1985).

Merican and Phillips (1985) found in their study of solid waste production from cage culture of rainbow trout that there was a considerable variation in the quantity of sedimenting material collected between cages, between farms and between cages and control sites. The amount of material collected below cages of each farm, but one, studied was significantly greater than the amount collected at the control site, showing a significant input of solids

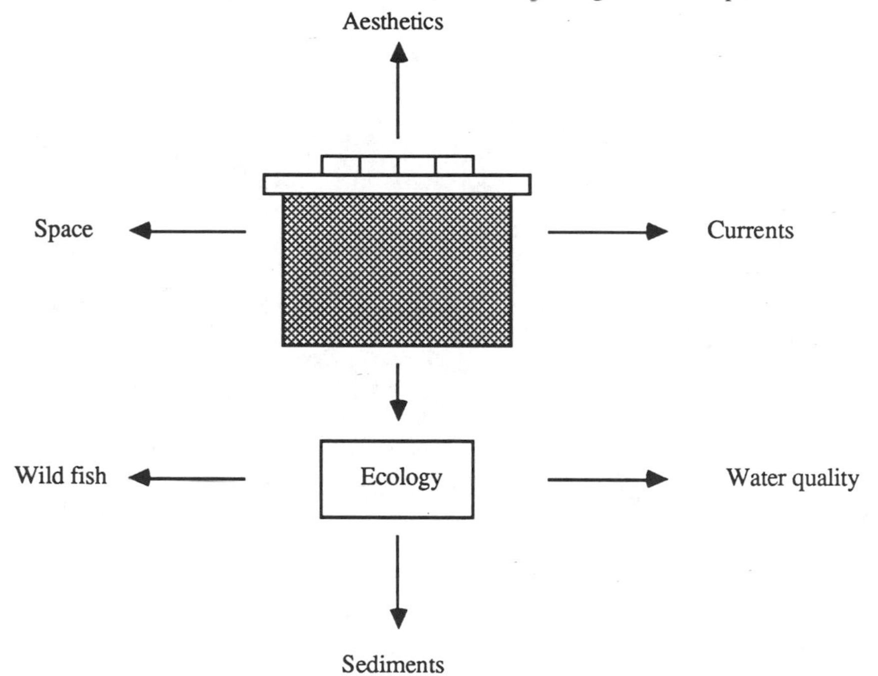

Fig. 11.3 Interactions between fish cages and the environment (from Phillips and Beveridge, 1986)

below the cages. The combined mean total sedimentation rate was 149.6 ($\pm$ 78.7) g in 24 h. The authors found that there was a significant correlation between waste output per unit biomass (i.e. grams per kilogram of fish, in 24 h) and feeding rate, but no significant correlation between output and fish weight, suggesting that feeding rate, rather than fish size, was an important determinant of solid waste production. They consider that, with food conversion ratios ranging from 1.4 : 1 to 3.2 : 1 (i.e. 1.4 t of food produces 1 t of fish), a solid waste output per tonne of fish produced is of the order of 0.54 t and 1.27 t.

According to Phillips (1984) the effect of cage farming on the benthos is usually noticeable in Scottish lochs and on the sites studied a large biomass of organic-pollution tolerant oligochaetes and chironomids was found immediately below and adjacent to the cages.

Nutrient enrichment causes increased algal production and in extreme cases eutrophication, causing algal 'blooms', has occurred. In Scottish lochs and reservoirs, likely to be sites for cage rearing of trout or young salmon, phosphorus is normally the critical factor which affects algal growth because of its scarcity. In northern Scotland the majority of standing waters are oligotrophic and relatively non-productive, with phosphorus levels well below 10 $\mu g\ l^{-1}$ or 10 $mg\ m^{-1}$ total P.

The method for predicting the effects of cages on phosphorus concentration is based on a model by Dillon and Rigler (1975) summarized in the equation:

$$[P] = Tw \cdot \frac{L}{\bar{Z}} \cdot (1-R)$$

where $[P]$ = predicted increase in phosphorus concentration in the loch caused by the cages ($mg\ m^{-1}$)

Tw = water residence or retention time (yr)

R = fraction of soluble phosphorus 'load' from cages sedimented and lost to the sediments

L = areal loading rate of soluble phosphorus ($mgP\ m^{-1}\ yr^{-1}$)

Z = mean depth (m)

Basic information on rainfall, catchment area size and loch volume can be used to estimte Tw. The fraction of phosphorus sedimented to the sediments has to be calculated theoretically from the water residence time and mean depth:

$$R = 0.426 \exp. \left(-0.271 \frac{Z}{\overline{Tw}}\right) + 0.574 \exp. \left(-0.009\ 49 \frac{Z}{\overline{Tw}}\right)$$

Phillips considers that a pessimistic estimate of the waste phosphorus discharged as soluble phosphorus into the water can be made as 50 per cent of the *total* loading with food conversion ratio of 2.0 : 1, that is 10 kg of soluble waste phosphorus per tonne of fish produced. Phillips and Beveridge (1986) calculated the budget for nitrogen and phosphorus for a Scottish rainbow trout cage farm (Fig 11.4). It showed that 85 per cent of the phosphorus and 80 per cent of the nitrogen fed to the trout on this particular farm was lost to the environment.

On the credit side it should be pointed out that cage culture in standing waters can be beneficial to the indigenous fish stocks. The effects of cage culture can affect the growth rate, abundance and survival of the indigenous fish. For example, dense populations of bluegill and redear sunfish have been recorded near channel catfish cages in the United States and many species of predatory and non-predatory fish have been caught in greater numbers adjacent to channel catfish and rainbow trout cage farms to which they are probably attracted by the waste pellets passing out of the cages. Phillips *et al.* (1985) have shown that rainbow trout cage culture improved the growth rate of roach, brown trout and stocked rainbow trout in some Scottish lochs and gave instances where the indigenous fish have been recording eating waste pellets.

However, the future use of lakes and reservoirs as sites for the cage rearing of fish will have to proceed with caution, and every new proposed site will have to be carefully monitored before cage rearing is permitted. In Norway, eutrophication of acidified lakes by cage farming is likely to be welcomed.

Fish farms have a number of other effects on populations of wild fish and the environment. These include the spread of disease and parasites and the effects on wild fish populations from fish escaping from the farms. However, these effects can work in both directions and farmed fish can contract diseases from wild fish and can also become infected from parasites emanating from the wild (Mills, 1990). An interesting example of this occurring is with the pearl mussel (*Margaritifera margaritifera* (L.)). Young salmon in recently established smolt-rearing units situated next to rivers in north-west Scotland where this mollusc exists have become heavily infected with their parasitic larvae known as glochidia. These glochidia can be released in very large numbers and attach themselves to the gills of fish where they develop into miniature mussels. Because the fish are held in the tanks in dense concentrations the risks of being infested are very great, consequently the losses of fish can be very high.

Reviewing the methods of fish culture and how closely they resemble natural aquatic systems can give one a very good insight into the successful management of fish farms and actions required to lessen their effects on the natural aquatic environment.

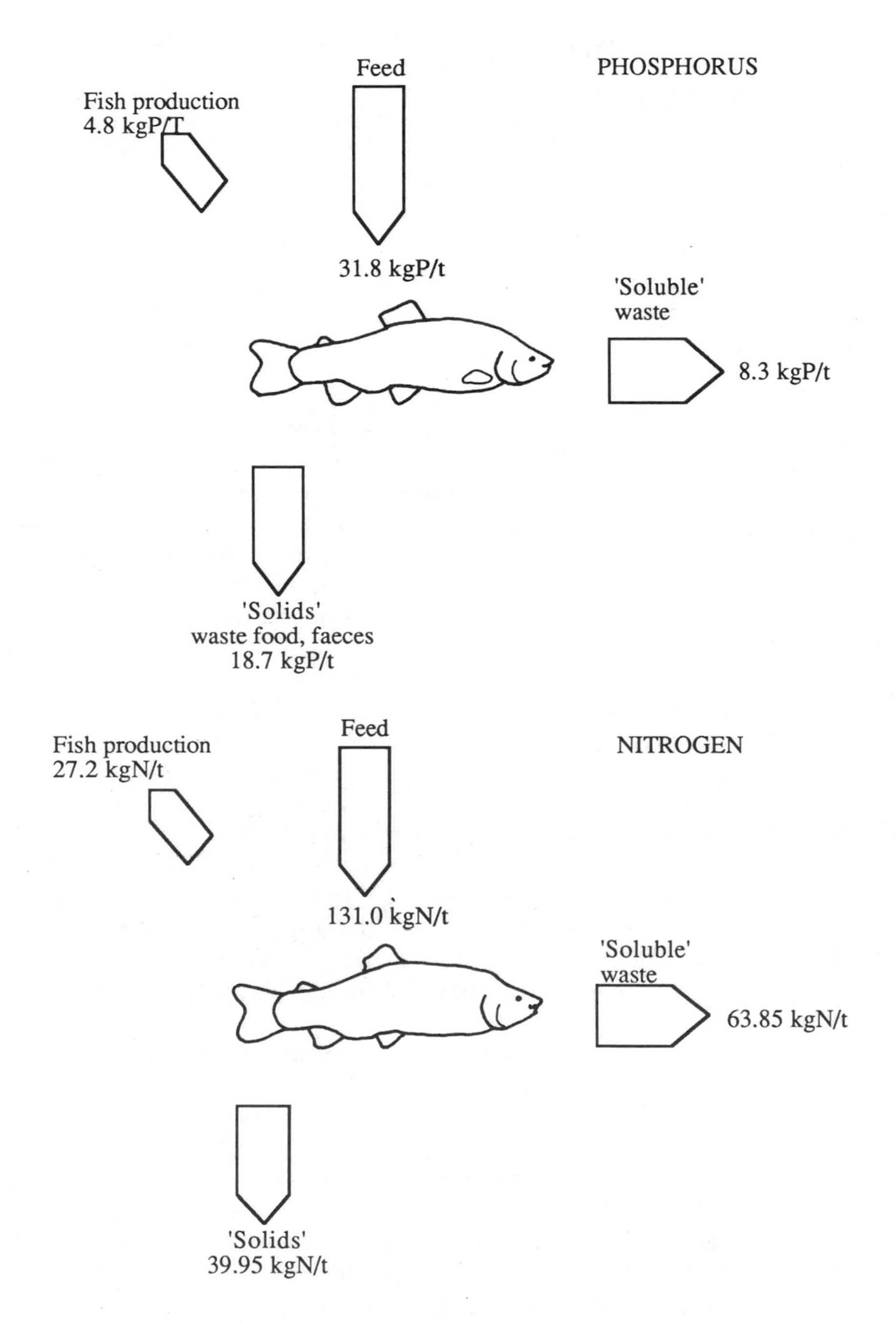

Fig. 11.4 Nitrogen and phosphorus budgets for a Scottish rainbow trout cage farm. (From Phillips and Beveridge, 1986)

12 Water abstraction and inter-river transfer

Water is a renewable resource and, as we have seen, is recycled through the hydrological cycle. However, because of our rainfall pattern only so much on average falls over any area every year but our demand for water rises each year as a result of population increase and industrial growth. The areas where the demand for water is most needed may change. For example, in the United Kingdom, in the recent past, there have been large demands in the midlands and southern England but, with the advent of the North Sea oil boom, the water demand in north-east Scotland, where the oil industry is based, increased very rapidly and water engineers had to look for sources of water to meet this demand. The Water Resources Board in its report in 1973 on *Water Resources in England and Wales* forecast a growth in demand on public supplies from 14 million cubic metres a day in 1971 to 28 million cubic metres a day by the year 2001. The sources of water and methods of transporting it to the public and industry are the responsibilities of the water engineers. In the United Kingdom generally speaking the areas of highest rainfall are in the west while the areas with the highest population density are in the east, so one is immediately faced with the problem of getting the water from where it is most plentiful to where it is most needed. This involves various methods of abstraction.

In large countries, such as China, the United States, Egypt and the Sudan the problems are much greater.

12.1 Methods of water abstraction

In any method employed for water abstraction there is some deleterious effect on the aquatic environment, however small. It is only where abstraction has a serious effect that measures have to be taken to minimize the problem. In the various abstraction methods listed the problems and solutions will be considered.

1. The simplest method of abstraction involves merely placing a pipe in a stream and diverting some of the water down the pipe under gravity to the supply point. This method is still used frequently in rural areas. Such a method has little effect on the aquatic environment.

2. Where a bigger community is to be supplied with water a lake is frequently tapped and, in order to control water levels, a sluice or dam is built at the lake outflow. On a migratory fish river system this may impede the movement of fish unless a fish pass is incorporated into the dam structure.

3. If there is no convenient lake or pond, reservoir formation is usually necessary. In its simplest form a small catchment in a hilly area may be trapped using a small dam, and the water diverted to supply with the exception of a compensation flow which is allowed to pass downstream to maintain the stream environment. In this type of scheme there are few environmental problems that effect the fishing interests, although there might be objections to such a scheme if the reservoir were to flood an area of scientific or archaeological interest.

4. The formation of a large reservoir often entails impounding the upper reaches of a river in an area of high rainfall and diverting the water directly to the demand area. This may mean that the water is piped a great distance and will not be returned to that river system. On the other hand the water may be used within the catchment area and is returned to the river downstream of the reservoir, but usually with a reduced quality. There are more environmental problems associated with this type of scheme and these can include: changes in river flow, water temperature and water quality; obstructions and delay to migratory fish, and flooding of the spawning grounds and nursery areas of fish.

5. A reservoir may be sited close to a river and kept full by pumping water directly from that river. The reservoir water may then be used to supply a neighbouring town, for example as Oxford is from Farmoor Reservoir, or to supply towns and villages quite distant from it, for example Grafham Water in East Anglia. If the supply is used close to the reservoir and effluent returned to the river, there may be little effect downstream, but where the water supply is used outside the catchment area there may be problems as with (4) above. A further set of problems arises if the water from the reservoir is pumped to towns upstream whose effluent then returns to the river so that some of it may be pumped into the reservoir; this recycling is likely to accelerate eutrophication of the reservoir.

6. The reservoir described in (4) is usually referred to as a *direct supply* reservoir. Another type of reservoir, which is of greater value to the river system on which it is sited, is a *regulating reservoir*. In this case

water is released from the reservoir during periods of low river flow and abstracted downstream at the point of demand. In other words, the river flow is regulated, flood flows are reduced to some extent as some of the flood water is stored in the reservoir, and drought flows are alleviated due to supplementation of the flow during dry weather. When water is not required for abstraction a compensation flow is discharged to maintain the 'health' of the river. Environmental problems still occur even with this type of reservoir and these include: unusually large fluctuations in river level which may result in rapid exposure and coverage of the river bed margins; a steady compensation flow which may be insufficient to scour the river bed of silt and weed growth and to assist the ascent of migratory fish; and reduced flows below the point of abstraction which can result in higher water temperatures, low dissolved oxygen levels, silt deposition and excessive weed growth.

7. A method of abstraction not requiring a reservoir is the pumping of water from the lower reaches of the river either directly into supply or into an existing storage reservoir. In this type of situation the river is left in its natural state to the 'very last minute' and water only abstracted after other water users (farmers, fishermen, etc.) have benefited from it. Such a scheme can only operate if the water is reasonably pure or where adequate water treatment facilities are available. An extension of this method is to pump the river water out through the neighbouring gravel by means of wellfields.

8. In chalk and limestone districts, water is held in the porous rocks and emerges as springs which feed the rivers. An inexpensive way of obtaining water for supply is to sink boreholes and pump away the water. By reducing or stopping the spring supply to nearby streams and rivers this may have catastrophic effects on their ecology. In recent years a system has been developed, mainly by the old Thames Conservancy, of pumping from boreholes but putting the water into the rivers and using these as aquaducts to carry water to the towns that need it, for example the rivers Lambourn and Kennet eventually supply water to metropolitan London. The pumping augments summer flows and so may be beneficial provided that the boreholes are not over-pumped, leading eventually to a decline in total river flows.

9. All the above abstraction schemes are relatively simple and involve single rivers. However, for the future demand in England and Wales a more integrated strategy of water resource development is required to meet the estimated deficit by the year 2001 and this is best achieved by *inter-river transfers* and river regulation. A number of inter-river transfer schemes have been proposed for England and Wales, some of them involving the transfer of water from as far apart as west Wales and the Thames. Two schemes at present in operation are the Kielder

scheme in north-east England, which involves the Rivers Tyne, Wear and Tees (Fig. 12.1) and the Ely Ouse scheme involving a complex of rivers flowing into the Cambridgeshire Ouse and the Stour and the Essex Blackwater. In the transfer of water from one river (the donor river) to another (the recipient river) there are likely to be a number of physical, chemical and biological effects in both rivers. A consideration of these effects will bring all our training in freshwater ecology into play and will serve as a useful revision of many aspects of this subject.

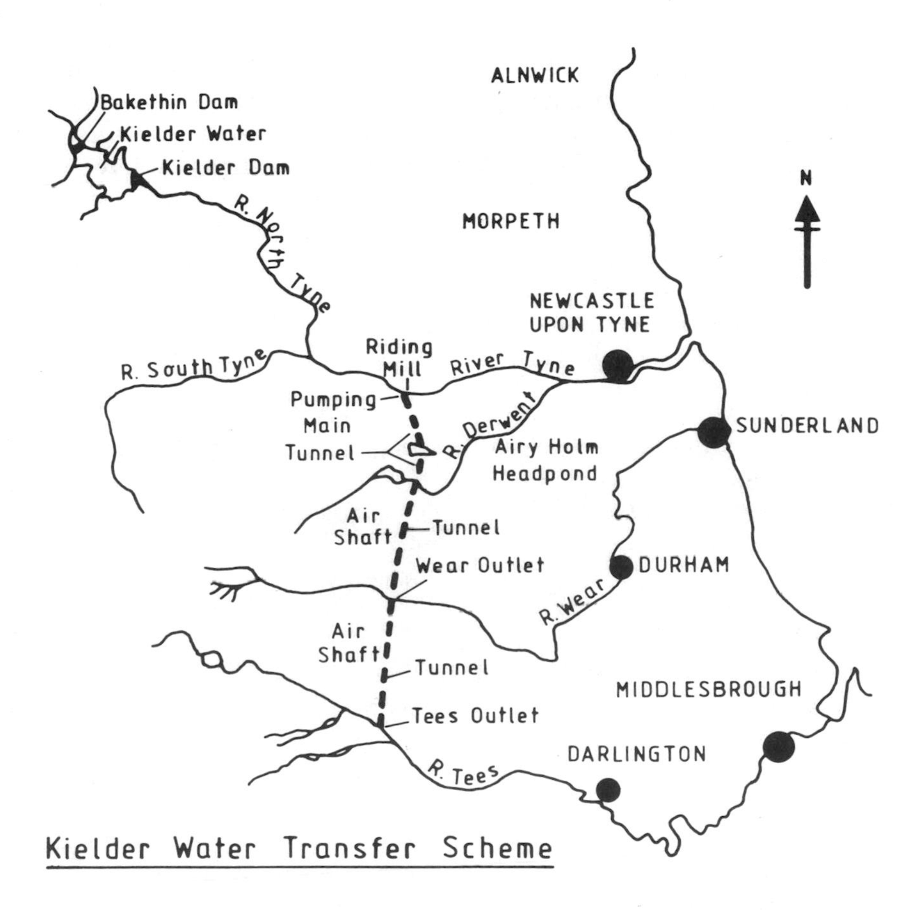

Fig. 12.1 Kielder Water Transfer Scheme

12.2 Effects of water transfer

12.2.1 Physical effects

CHANGES IN FLOW REGIME

In some cases the flow transferred from the donor river to the recipient river will amount to several times the average daily flow of the recipient river and therefore bring about conditions associated with flood flows. This could lead to the need for channel modification.

Effects of channel modification. Enlarging of the recipient river channel could result in the destruction of a fishery or, alternatively, the improvement of an existing fishery. As we have seen, fish and their food organisms require certain depths and cover. Fish will not remain where there is insufficient depth of water or inadequate cover. A straight, trapezoidal, evenly graded channel, steep-banked and devoid of cover, presents the maximum of unsuitability for all plant and animal life. On the other hand, a meandering channel cut with one or more lateral shelves (known in engineering terms as *berms*) below water level, on at least one side, depended on the bends, and with natural cover left intact, will rehabilitate as natural scour and bed reorganization take place, and aquatic vegetation re-establishes itself on both the river bed and river banks (Figs. 12.2 to 12.7). The lateral shelves will support the major invertebrate communities and the deeper areas will provide adequate depth for fish at all river flows.

Other effects of channel modification will be high suspended solid levels during excavation works and for such time afterwards as the new channel scours. This is likely to have a damaging effect on the aquatic flora and the invertebrate and fish fauna through discoloration of the water and sedimentation, especially in pools. Furthermore, sedimentation over the spawning gravels of salmonid fish would reduce the hatching success of the eggs incubating in the substrate.

Effects of seasonal transfers. On recipient rivers one effect of seasonal transfers will be to maintain flows in dry weather periods at levels above those previously occurring. This is likely to be beneficial to the aquatic fauna and to fisheries and anglers.

The rate at which a transferred flow is introduced into, and arrested from a recipient river is important. Sudden deepening of the river may also endanger people in or near the river, while a sudden reduction in flow may lead to stranding of fish in shallow pools and the death of invertebrates left 'high and dry' when the river level falls.

Fluctuations in transfer volumes into migratory fish rivers could be beneficial, provided that the changes in flow are not abrupt. Seasonal dry weather transfers into rivers will tend to even out the annual flow pattern

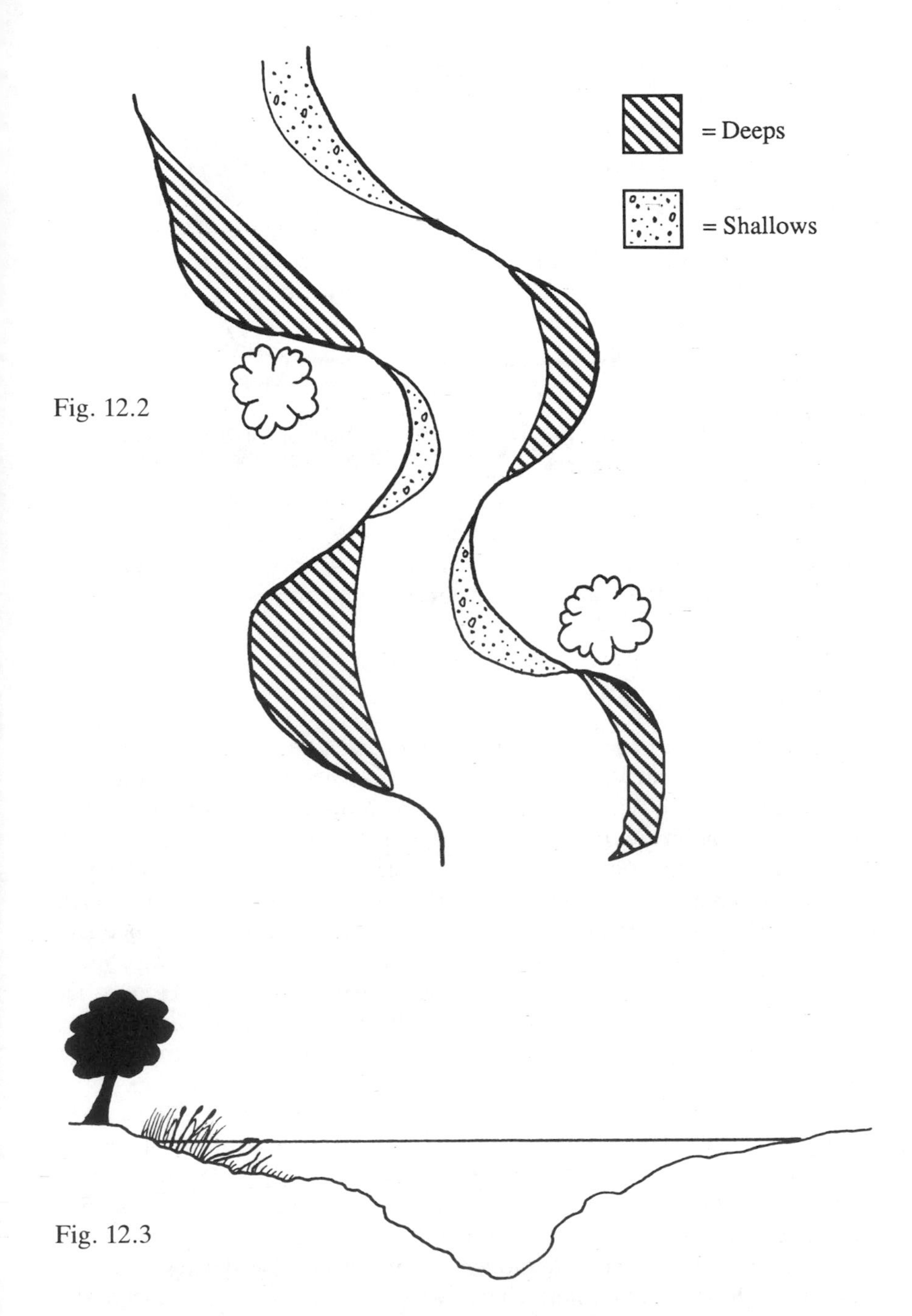

Fig. 12.2 and 12.3 A plan and profile of a natural river which is ideal for fish and angling, with adequate cover and food, but inadequate for water transfer.

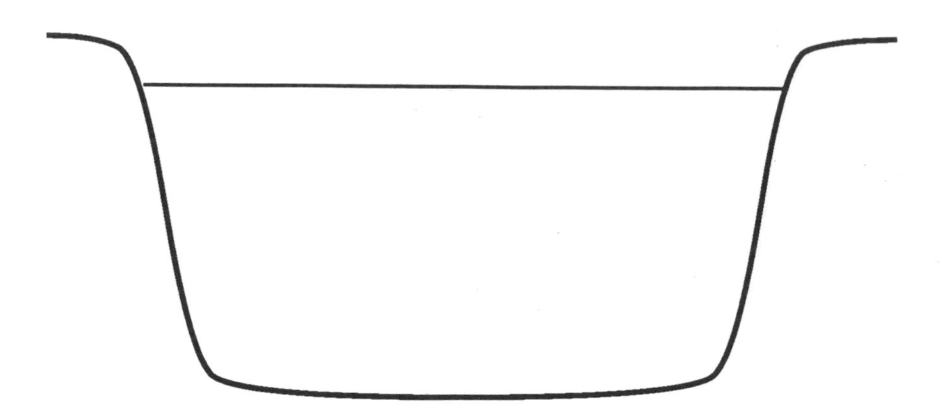

Fig. 12.4 A profile of a dredged river giving a uniform channel all of one depth and width with a uniform gradient, making it ideal for water transfer but disastrous for fish.

towards that characteristic of chalk streams. This would tend to preserve the 'wetted perimeter' and thus the invertebrate fauna, as well as the depth and flow.

CHANGES IN FLOW VELOCITY

These will affect the river bed stability and sedimentation, survival of rooted plants and invertebrates and the suitability of the habitat for various species of fish at various stages in their life history. Flow velocity is important in relation to the ability of rooted aquatic vegetation to maintain itself, as it can cause scouring of the roots and damage by buffeting in the current. The type of vegetation is likely to change with increased flow velocity especially when this results in changes in the depth of the channel, the nature of the bed and the amount of turbulence. Flow velocities may affect the invertebrate fauna through the shifting of the bed material, leading to destruction of the fauna, or its occlusion by siltation; downstream displacement of certain species; alteration in the texture of the river bed and therefore a change in species composition, favouring those with higher velocity preferences or, in the donor river, those with lower velocity preferences; destruction or increase in rooted aquatic vegetation leading to a reduction or an increase in the abundance of invertebrates, and a removal of or increase in detritus leading to the removal of, or establishment, of detritus feeders. The effects on fish may include the scouring of salmon and trout redds; the washing away of the newly hatched alevins, and the movement of cyprinid fish species to slower flowing reaches in the recipient river and the spread of cyprinid fish in the donor river as a consequence of reduced flows.

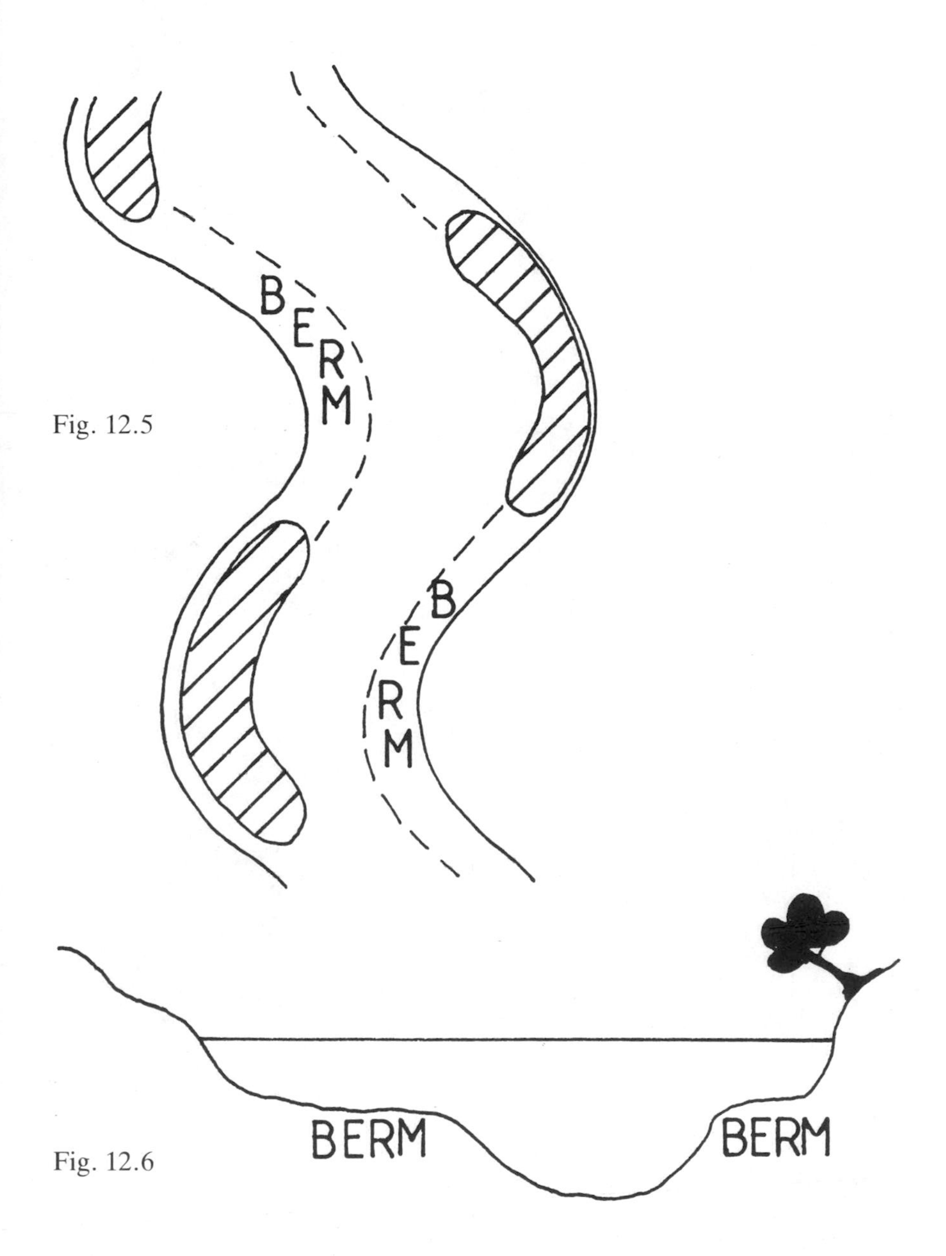

Fig 12.5 and 12.6 Plan and profile of a modified river to produce a compromise giving an efficient channel for water conveyance and a sinuous channel of varying depth for fish cover. The berms provide a suitable environment for plants and food organisms, while trees and bushes are left intact or replanted for cover.

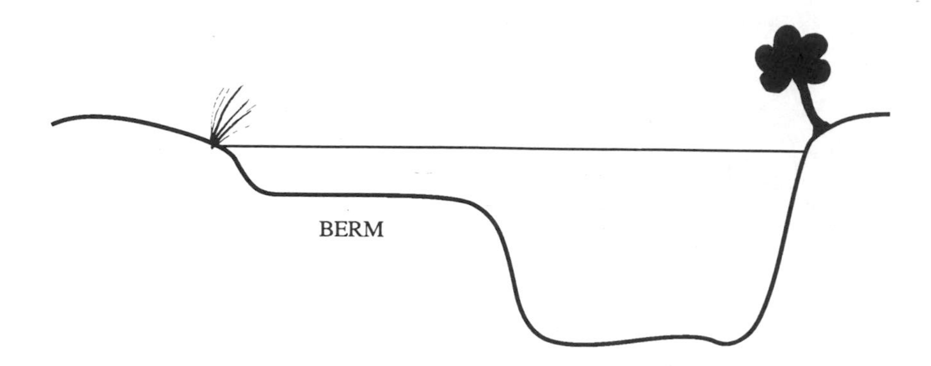

Fig. 12.7 A modified river showing provision for extra depth on bends with a berm on only one side.

CHANGES IN SUSPENDED SOLID LOADS AND TURBIDITY

Suspended solid loads may increase as a result of bank erosion. The increased turbidity may smother the fauna on the river bed and also reduce the light penetration which could affect primary production.

CHANGES IN WATER TEMPERATURE

A river which is receiving water continuously from a reservoir may experience less extreme temperatures than before provided the water in the reservoir is not stratified and that the water is not drawn from the hypolimnion. It was found in the Saskatchewan River in Canada that the kinds and numbers of Ephemeroptera and other insects were greatly reduced downstream of a reservoir. This was attributed to changes in river temperature caused by the reservoir. The river is now warmer in winter and cooler in summer. Consequently, the mayflies and other insects with strict thermal requirements could not hatch and grow successfully. The effect was evident 112 km downstream.

Thermal changes downstream of a reservoir are also known to affect fish. Anglers on the Severn maintained that their catches were less when stored water was released upstream from the regulating reservoir Lake Vyrnwy. A study of the effect of releases from Kielder Reservoir was made by the Freshwater Biological Association (Crisp, 1984) and a change in the water temperature regime was noted. It is believed that this temperature change will affect the upper 10 km of the River North Tyne immediately downstream of the reservoir. The principal effect has been to make the North Tyne warmer in late autumn and cooler in spring and early summer. It is suggested that an increase in autumn temperatures by several degrees would lead salmon ova to hatch 50–60 days earlier than previously. Lower temperatures in spring mean that it might take a further 40 days before the temperature rises to a level (7 °C) at which salmon fry start to feed. Thus ready-to-feed fry would emerge when water temperatures were too low for them to feed and they could consequently starve and die. Cave (1985) points out that as smolts usually start to migrate when water temperatures rise above 10 °C, the changed water temperature regime in the North Tyne could delay migration by as much as 40 days and smolts could miss the limited time they have to pass through the estuary to the sea. This time, referred to as the 'window' period, is all the more noticeable in the Tyne estuary where dissolved oxygen levels drop after the spring floods have passed to less than 40 per cent in some reaches and thus prevent smolt movement.

In addition to the effects on the recipient river there could be drastic changes to the donor river. The dramatic effects this could have are best illustrated in the plans for diversion of water from the Chang Jiang, the largest river in China to the drought-stricken North China Plain. The initial plan involves pumping 1000 m^3 of water a second and transferring 30 km^3/year from the Chang Jiang. The effects this would have on the aquatic environment are dealt with very fully by Jinghua and Yongke (1983) and Xuefang (1983).

12.2.2 Chemical effects

CHANGES IN HARDNESS

River transfers in England and Wales will tend to run from north-west to south and south-east. This will mean that soft water will tend to flow into hard water or hard water will flow into hard water. In most hard waters the calcium concentration is well above the limiting level for hard water invertebrate species and even a considerable dilution with soft water will not have a great effect on the invertebrate fauna. However, there is a range of alkalinity values, between 10 and 75 mg l^{-1} $CaCO_3$, in which the photosynthetic activity of dense plant growth or algal blooms can cause the pH value of the water to rise to levels lethal to fish during the daytime. This situation could arise, as a result of water transfer, in the sluggish stretches of a donor river downstream of the abstraction point.

CHANGES IN INORGANIC NUTRIENT CONTENT

The transfer of rich inorganic salts to a nutrient-deficient stream is likely to increase primary production. However, if the process went too far it could result in excessive growths of *Cladophora*.

CHANGES IN POLLUTANTS

Changes in the pollutional qualities of the recipient water can occur if the composition of the transferred water affects the toxicity of the pollutants already present in the recipient river. Changes in pH may occur which could increase the toxicity of those poisons – ammonium and metal cyanide complexes – whose toxicities are dependent on their degree of ionization. In general, addition of extra quantities of clean water will have a beneficial effect in that the dilution of the effluents discharged to it will be increased.

12.2.3 Biological effects

TRANSFER OF FISH DISEASES AND PARASITES

The transfer of such diseases as the salmon disease, UDN (ulcerative dermal necrosis) and the larval stages of the eye fluke parasite (*Diplostomum*) are both likely. However, it is not known in the case of the latter what the effect of such a transfer would be.

TRANSFER OF PHYTOPLANKTON AND MACROPHYTES

Addition of water to the recipient river already containing a heavy growth of algal cells may have significant effects, particularly when the introduction is made into the upper reaches from a reservoir or from the lower reaches of the donor river, by the prolongation of the time available for reproduction by the algae. Normally phytoplankton do not develop to severe 'bloom' proportions in northern European rivers because their generation time is a few days and the discharge time of even slow rivers is only of the order of a few weeks. Thus only relatively few algal generations occur during transit from source to mouth. Even so there is a large enough population in the lower reaches of some slow rivers to cause problems to water boards abstracting near the sea. Problems of this nature have occurred on the Ely Ouse transfer scheme.

The disadvantageous effect for fisheries of excessive algal and macrophyte growth lies in their effects on the toxicity of polluting conditions through their ability to increase pH values and reduce CO_2 concentrations in daylight and to reduce dissolved oxygen levels by respiration during hours of darkness, and by decomposition, and, in the case of algae, the turbidity produced which is not aesthetically pleasing.

TRANSFER OF INVERTEBRATES

The transference of parasite-bearing invertebrates, such as *Gammarus*, *Asellus* and *Lymnaea*, will spread the incidence of internal parasites such as *Acanthocephala* and *Diplostomum*.

TRANSFER OF FISH

The transfer of young salmonid fish could be an advantage to the recipient river, but on the other hand the introduction of cyprinid species or predatory fish to a salmonid river could be detrimental.

12.2.4 Effects on reservoirs

THERMAL STRATIFICATION

This layering of water could cause difficulty to the water engineer over the siting of the water draw-off point to ensure the best quality water for abstraction or to prevent poor quality water being released into the rivers. For example, the bottom water may be high in humic acids and oxides of iron and manganese.

WATER QUALITY

The quality of water being introduced to the reservoir will be of importance, and nitrate and phosphate loads, from such sources as agriculture and forestry, would need to be low to prevent eutrophication.

SHORE EROSION

Shore erosion due to the drying-out effects from draw-down and fluctuating water level and wave action could seriously affect productivity. This erosion also leads to the loss of some of the littoral invertebrate communities.

INTRODUCTION OF FISH

Fish may be inadvertently introduced to the reservoir by being pumped up from the river in their egg or larval stages if the reservoir is partially refilled in this way.

Fortunately most of the environmental problems associated with water abstraction schemes never materialize, at least not in the United Kingdom, as before any scheme is allowed to proceed the appropriate water authorities commission consultants to undertake an environmental impact assessment. Subsequent reports include recommendations for overcoming or preventing many of these problems. In addition, the natural process of democracy provides for objectors to voice their opinions at public hearings after which the relevant government minister gives his decision. Some areas are also protected by being given the status of sites of scientific interest and two of the largest rivers in the United Kingdom, the Wye and the Tweed, have been given the status of Sites of Special Scientific Interest by the Nature Conservancy Council. However, we all need water and it is in the interest of the whole community that conflicts between conservationists and developers are resolved sensibly without either the conservationists delaying needful developments and making unreasonable claims on the public purse or water bodies riding roughshod over river proprietors, anglers, conservationists and the like.

13 Hydroelectric development

Over the years many fast-flowing rivers in areas of high rainfall have been progressively harnessed to produce energy from the stored water power. These hydroelectric schemes have proliferated in both temperate and tropical countries. In the former, hydroelectric schemes have been widely developed in Canada, France, Iceland, Ireland, New Zealand, Norway, Scotland, Spain, Sweden and the United States. In tropical areas massive schemes exist on the Nile, Zambezi and Orange Rivers in Africa, while similar large developments exist in Australia, India, Nepal and South America. In temperate regions these schemes have tended, with some exceptions in Canada, to affect chiefly the migratory fish stocks of the rivers but in tropical regions the ecology of a wide range of fauna, including man, has been disturbed (Lowe-McConnell, 1966).

13.1 Types of hydro scheme

Hydroelectric installations are of four main types: (1) a simple dam or barrage which diverts water for use in an adjoining watercourse or in another catchment area; (2) a dam, normally larger than the diversion dam, which impounds water which is piped to a power station either incorporated in the dam or located some distance downstream of the dam and, after passing through the turbines in the station, is returned to the river. These first two types are sometimes known as 'run-of-the-river' schemes and often the generating station is built into the dam; (3) a type involving a storage reservoir some distance from, and usually at a much higher level than, the power station, thus providing a greater 'head' of water so that less water is required to produce the same or greater amount of electricity as in type (2); (4) a pumped-storage scheme involving two reservoirs at different levels (e.g. Loch Awe and Loch Cruachan in Scotland), so that during the day at periods of peak demand water is released from the upper to the lower reservoir via a high-head power station, whereas at night, when there is spare cheap power, water is pumped back up to the top reservoir to repeat the process the following day.

Plate 27 A hydroelectric reservoir and dam in Ross-shire, Scotland. The Orrin Dam depicted here is part of the North of Scotland Hydro-electric Board's Conon Valley scheme. (Photo: North of Scotland Hydro-electric Board)

13.2 Effects on the aquatic and riparian habitat

Building such large structures across a river and impounding water behind them inevitably alters the natural regime of a river in much the same way as in the water abstraction schemes described in the previous chapter and many of the effects are the same but there are differences, many of them involving the structures themselves.

13.3 Hydro schemes in temperate regions

In temperate countries most hydro schemes are built on fast-flowing rivers originating in mountainous terrain and these are typically the sort of rivers frequented by migratory fish such as salmon and trout. So any barriers built across the rivers are obviously going to have some effect on their movements and well-being.

13.3.1 Obstruction to fish migration

During periods of low discharge the ascent of fish from the estuary may be delayed and fish in the river may be deterred from moving upstream. The question which arises is how long can delay in salt water or fresh water occur without affecting the fishes' ability to move upstream and to spawn effectively? There may be a change in the quality of the fresh water because of reduced flow, increased temperature and perhaps even pollution. McGrath and Murphy (1965) attribute a high mortality of adult salmon in the River Lee, Ireland, partly to a change in the quality of the water discharged from Inischarra Dam, as the dissolved oxygen in the deeper parts of the reservoir had reached a dangerously low level. This was a transitory condition and arose because of the decay of organic matter in the reservoir basin following the first flooding with water; the situation had been aggravated by the fact that the turbines had not gone into operation for some time after the reservoir was filled, which permitted the quality of the water to deteriorate. It has been shown that the deterioration of the water quality in the Saint John River, New Brunswick, has been exacerbated both by pollution loads and by a reduction in the biological assimilative capacity of the river as a result of hydro-power developments on the river.

On the Pacific coast of North America the five species of Pacific salmon enter the rivers in very large numbers, particularly the sockeye (*Oncorhynchus nerka* Walbaum) and and chinook (*O. tshawytscha* Walbaum), and as they may have to swim some hundreds of miles to their spawning grounds it is essential that their migration is not delayed for even a

Plate 28 An Atlantic salmon makes a vain attempt to leap the Domtar Barrage at Donnacona on the Jacques-Cartier River, Quebec. (Photo: M. Frenette)

day, as otherwise their reserves of energy may have been exhausted before they reach their destination. Fish passage facilities may not always be adequate to cope with large numbers of fish, particularly at high dams. It is therefore sometimes necessary to trap them below the dam and truck them upstream to a release point above the reservoir (Ebel, 1985).

Because salmon have a very strong 'homing instinct' and return to their parent river to spawn, the diversion of water from one river to another may result in the 'straying' of fish to a 'foreign' river into which their 'home' river water is being diverted.

Fish in the river below a power station may be subjected to a wide range of flows a number of times in a day. On the River Conon in Ross-shire, Scotland, the water level may rise to flood conditions and fall away to low summer flows as often as three times in the 24 hours. These rapidly fluctuating water levels may result in the exposure and drying out of redds constructed by the spawning fish along river margins at high flows and may result in large losses of juvenile fish stranded as flows are reduced and water levels drop.

Young fish, particularly seaward migrating salmon smolts, may be delayed in the reservoirs either through not finding the entrance to the statutorily required fish pass at the dam face or through being diverted into the power station tunnel. The numbers of young fish that are held back in the reservoir are frequently decimated by predatory fish and birds (Mills, 1964). Smolts can pass unharmed through large turbines with a head of between 30 and 40 metres. For higher heads they have to be excluded from the tunnel to the power station by small-meshed screens or other devices such as curtains of bubbles.

It has also been found that Atlantic salmon and sea trout that have spawned, when they are known as kelts, may also be delayed by dams and may take some time to descend the fish pass.

Frequently large spawning and nursery areas are either eliminated by reservoir formation or are reduced by diversion of their water supply to a neighbouring reservoir. There may also be some damage to other areas through rapid fluctuations in water level below a power station, resulting in exposure of redds during periods of frost and stranding both of spawning adults and of young fish. In addition, reduced flows may cut down the area available to fish (Mills, 1989a).

13.3.2 Effects downstream of reservoir

Water discharged from reservoirs is generally free of coarse sediments, as the reservoir will trap nearly all of the inflowing coarse material. In New Zealand the Matahina Reservoir (Callander and Duder 1979) and impounded Lake Roxburgh (Jowett and Hicks, 1981) were found to trap 70 per cent and 80 per cent of the incoming sediment respectively. In Lake

Roxburgh the median concentration of inflowing suspended sediment over an 18-month period was 15g/m^3 while the mean suspended sediment concentration of water flowing out of the lake was only 5g/m^3. There is therefore a tendency for degradation to occur downstream as the river attempts to regain a balance between discharge slope, sediment transport and grain size (Irvine and Jowett, 1987).

Morphological changes in the river may occur as a result of the changes in the flow regime. An increase in flow can cause erosion of river banks and result in increasing the mean size of particles remaining. This can lead to decreased turbidity which will therefore benefit primary production and so the number of invertebrates and consequently the growth of fish. On the other hand, reduced flows can result in sedimentation with decreasing substrate diversity. This can reduce the amount of algae which will have a deleterious effect on the invertebrate (Ward, 1976) and fish fauna. Silt accumulation in the Lower Waitaki River bed in New Zealand has been attributed to the increasing amount of flow regulation by upstream hydro-electric storage reservoirs (Jowett, 1983).

It has been found in New Zealand that the riverine bird fauna has also been susceptible to the impact of hydro-electric dam construction with subsequent changes to the downstream flow regime (Hughey, 1987) and certain of the less common species of riverine bird could be endangered.

13.3.3 Effects of impoundment on lakes and reservoirs

After impoundment, resulting either in reservoir formation or in raising the level of an existing lake, there is usually an increase in the general biological productivity with a risk of severe eutrophication due to leaching of nutrients from newly inundated ground. Frequently the production of planktonic and semi-planktonic crustaceans increases as a result of this nutrient leaching and decaying terrestrial vegetation, and some of the important food sources for fish during the early impoundment period will be terrestrial invertebrates such as earthworms and terrestrial insect larvae. The forage area of fish is usually greatly increased and therefore their population density is temporarily reduced. Immediately after impoundment the growth rates of fish such as trout, charr and perch have been shown to improve substantially, but in almost all cases it has been found that these improved growth rates only last for a few years. However, if the area of shallow water is increased after impoundment and water levels remain relatively stable the initial increase in fish production may be sustained; Campbell (1963) suggested that this might be the case with trout in Loch Garry, Inverness-shire, Scotland.

Elder (1966) pointed out the effects of fluctuating water levels on the littoral vegetation and fauna. Those species closely associated with the littoral vegetation will be greatly reduced or even eliminated, notably many of the larger insect larvae and crustaceans, while chironomid larvae may increase. However, in terms of the production of fish food the net result of these changes is loss of the most valuable food organisms and survival of types which tend to be less available to fish. The effects of water level fluctuations on the invertebrate fauna have been most fully studied in Swedish hydroelectric reservoirs. It was found that the fluctuations in water levels caused a quantitative reduction in the bottom fauna amounting to 75 per cent in the zone of water level fluctuation and 25 per cent in the remaining area. Maximum abundance of the invertebrate fauna occurred immediately below the limit of draw-down (Lindstrom, 1973). After decades of regulation, the shrimp (*Gammarus lacustris Sars*) has been reported to have disappeared from some severely regulated lakes in northern Sweden (Grimås, 1962).

In some reservoirs in nothern Sweden there have been changes in the aquatic vegetation caused by the reduced yearly water level amplitude which has given rise to a considerable practical and financial problem as the large increase in *Sparganium* species with long floating leaves interferes severely with timber-floating operations and fishing with nets.

The breeding success of certain fish species such as perch (*Perca fluviatilis* L.), roach (*Rutilus rutilus* L.) and pike (*Esox lucius* L.), which spawn among submerged vegetation in shallow water, may be severely impaired by water level fluctuations. Other fish species, which lay their eggs in gravel, have been found to extend their spawning ranges to take advantage of suitable gravel uncovered by erosion when water levels drop. The charr (*Salvelinus alpinus* L.) in Sweden and the lake trout (*Salvelinus namaycush* (Walbaum)) in Canada were not eradicated due to their eggs being stranded by falling lake levels, but extended their range into deeper water (Elder, 1966).

As a result of impoundment there may also be changes in the fish habitat leading to population changes. For example, as a result of the construction of the Tennessee Valley Authority's Melton Hill Dam on the Clinch River the proportion of game fishes such as white bass (*Morone chrysops* (Rafinesque)) and gizzard shad (*Dorosoma cepedianum* (Lesueur)) increased. Certain species of fish (e.g. largemouth bass, *Micropterus salmoides* (Lacepède), goldfish, *Carassius auratus* (L.) and mosquito fish, *Gambusia affinis* (Baird and Girard) only appeared after impoundment (Fitz, 1968)).

After the construction of diversions and pipeline connections between water bodies or different catchments there is the danger of accidental introductions of new fish species. It is believed that the mormyrid fish (*Mormyrus macrocephalus* Worthington) gained access to Lake Victoria when the Ripon Falls were flooded by the completion of the Owen Falls dam further down the Nile. The Nile perch (*Lates niloticus* (L.)) also gained access to Lake Victoria from the Nile, possibly by passing through the turbines of the Owen Falls Dam during a certain stage of their operation.

13.3.4 Effects of pumped-storage

There may be some limnological effects where there are wide ranges in the water levels of both reservoirs and some stocks of indigenous fish may be affected. However, Smith *et al.* (1981) in a study of Loch Awe and Loch Ness, Scotland, on whose shores pumped-storage schemes operate, found negligible effects on the zooplankton and indigenous fish. They also found that growth rates and life cycles of species inhabiting these two lochs appear to be unaffected by the schemes. The authors did suggest that fluctuating water levels in Loch Awe caused by the pumped-storage scheme might favour *Gammarus pulex* more than *G. lacustris*. They also recorded the occurrence of the lamellibranch *Pisidium personatum* Malm from the man-made upper reservoir, Cruachan, on the Awe scheme, which could only have originated from either Loch Awe or one of its tributaries. Arctic charr originating in Loch Awe also colonized Cruachan. Kaatra and Simola (1985) describe the impacts of water level regulation of Lake Inari in northern Finland and describe some of the compensation measures employed including stocking the lake with fish.

13.3.5 Major environmental impacts

Hydroelectric schemes can have much wider-ranging effects than listed above and some schemes in Canada could have mind-boggling repercussions for aquatic environments incredible distances from the actual dam or reservoir. One of these is the James Bay project in northern Quebec involving the diversion of whole watersheds and a lake of nearly 8000 km^2. However, it is the impact that the Bennett Dam, built at the outfall of Williston Lake in the Rocky Mountains, had on the Peace–Athabasca Delta 1126 km downstream in Alberta which is most sobering. Low water levels on the delta coincided with the upstream filling of Williston Lake during 1968–70, exposing thousands of hectares of marsh and lake bottom, advancing plant succession (towards the less desirable willow communities), threatening access of migrating fish to the delta's spawning lakes, and reducing nesting habitat for waterfowl and overwinter habitat for muskrats. The resultant drastic decline in the muskrat population affected the economy of Fort Chipewyan and the lifestyle of its native people because of their dependency on trapping. The long-term effects of the changed regime were predicted to decrease the important shoreline of perched basins by about 50 per cent, to cause plant succession to proceed uninterrupted for longer periods of time, thereby accelerating the ageing of the delta, to shift plant zones to lower elevations along lake margins, and to reduce the vertical ranges and area of early successional plant communities by as much as 50 per cent. Waterfowl production was expected to decline by 20–25 per cent, and muskrat populations by 40–65 per cent. Bison grazing meadows were expected to suffer losses although moose habitat was expected to improve. The decrease in water levels would increase the risk of delay in the spawning runs of the walleye to the delta lakes. Fortunately remedial measures have been implemented since this study in 1973 (Townsend, 1975).

In Europe, hydroelectric developments are also having wide-ranging effects on some of the major river systems. The River Danube, once running unhindered through some of the most spectacular scenery in central Europe to its delta and the Black Sea, has now, along with many of its tributaries, been dammed for navigation, water consumption and energy production (Bacalbasa-Dobrovici, 1985).

13.4 Hydro schemes in tropical regions

Some of the problems arising from the construction of high dams on the Zambezi at Kariba and Cabora Bassa and the Nile at Aswan involve the river environment as a result of the consequent regulated flows.

Plate 29 The Owen Falls Dam on the Victoria Nile. (Photo: D. Downie)

13.4.1 *Effects downstream of reservoir*

PROBLEMS OF SILT-FREE WATER

The silt load brought down to a reservoir by the inflowing river settles out behind the dam, and in some parts of the world, where the amount of suspended material carried downstream by rivers is very high, reservoirs may have a very short life as a result of becoming filled with silt. In large reservoirs engineers design the dam so that there is a 'dead storage area' included in the carrying capacity of the reservoir to cope with the high silt loads. As a consequence the water released from the reservoir is devoid of silt and due to the physical characteristics of silt-free water it becomes an active agent of erosion. The so-called Nile-cascade scheme for building a series of barrages between Aswan and Cairo aims at protecting the downstream Nile channel against the erosive action of the silt-free water. The reduction of the Nile's annual flood load of sediments that the natural river system brought to its delta is causing serious marine erosion and an alarming retreat of the delta shoreline, resulting in an upstream penetration of salt water with a loss of irrigable areas (Kassas, 1974).

The trapping of silt by Lake Kariba has resulted in an impoverishment of the alluvial flood plain in the Mana Pools area some 80–100 km below the dam. The Zambezi River usually rises in December with a peak in March to April and it is in flood for two or three months or even longer. Formerly silt loads were swept across the flood plains with the floods and, on the gradual recession of the water, regeneration of the vegetation occurred with great vigour, but since the building of the Kariba Dam the silt load and the degree of regeneration have decreased. The Zambezi was the main contributor of this rich deposit of silt as its tributaries Kafue and Luangwe contributed little (Fig. 13.1). The flood plains of the Mana Pools area are of great importance to the conservation of wildlife in the Zambezi valley and during the dry season they are the main grazing areas. Under the present-day regime they are occupied by the large mammals from May to November, but were used for a shorter period in the pre-impoundment days due to the area being flooded. The species composition of the grass sward in the Mana Pools area is declining and being replaced by unpalatable herbs. There is little regeneration of the woody fodder species and over-utilization by the animals of what remains.

REDUCTION IN NUTRIENT FLOW

The nutrients associated with a rich silt load are also retained within the reservoir so that the released water is usually devoid of many nutrients. This may have a profound effect on fisheries downstream of the dam and a classical example of this lack of nutrient flow has been the collapse of the sardine fishery off the Nile Delta. Prior to 1964 18 000 tonnes of sardines

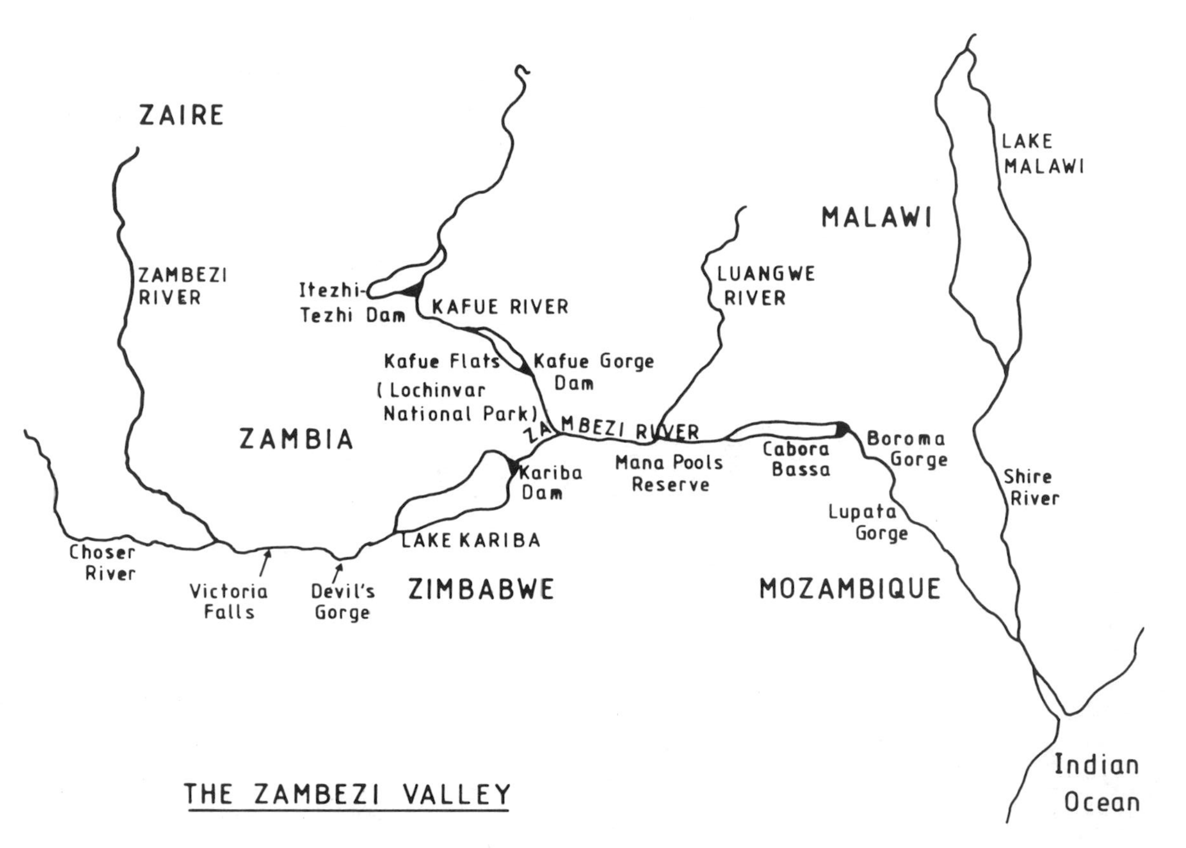

Fig 13.1 The development of hydroelectric power on the Zambezi River.

were landed annually but after 1969 the catch declined to only 500 tonnes (Abu-Zeid, 1983).

EFFECTS OF RELEASES AT ABNORMAL SEASONS

Releases of large volumes of water from the reservoir during the dry season may well produce flows much greater than the mean dry season flow of the pre-impoundment era. This will in effect cause unseasonal flooding at a time when many animals are particularly vulnerable to such catastrophes. For example, floods may flush out backwaters and small pools in which amphibia are breeding; sandbanks adjacent to the water's edge and in which reptiles have laid their eggs and on which some wading birds have nested may be inundated, and bank-nesting birds such as bee-eaters may have their nests flooded or the banks themselves may collapse.

EFFECTS OF A LACK OF A WET SEASON FLOOD

If there is no wet season flood due to the flood waters being impounded in the reservoir there is likely to be an impoverishment of the animal food supply. Baboons, for example, work industriously over the hard compact clay sections of drying river areas to obtain the molluscs, insects, amphibians and small reptiles which seek shelter in the crevices as the areas dessiccate. Each flood would normally serve to replenish the productivity of such areas. Without adequate flooding of the plain it becomes available earlier to the fauna moving from their wet season dispersal areas or it will be available to them all the time. The floodplain will thus not be rested normally and some species, such as waterbuck, will hardly move off it at all (Atwell, 1970). The pools in depressions and oxbows will not be flushed out and will therefore become choked with aquatic plants such as the water fern (*Salvinia*) and water lettuce (*Pistia*) thus limiting the space for amphibia, reptiles and birds.

13.4.2 Effects of impoundment on the reservoir

One of the major problems affecting newly formed reservoirs in parts of the tropics is excessive weed growth and vast mats of species such as the water hyacinth (*Eichhornia crassipes*), water lettuce (*Pistia stratiotes*) and water fern (*Salvinia auriculata* Aublet) may cover large areas. On Lake Kariba mats of water fern covered an area of over 1000 km^2.

One of the problems most affecting man is the high incidence of the snail-borne disease bilharzia due to the massive increase in the numbers of water snails.

The fish species composition in the reservoir will change to some extent as river species are gradually replaced by still water species. However, the distribution of species in the reservoir may initially be affected by limnological conditions in the reservoir. For example, during and after the filling of the Brokopondo Reservoir in Surinam there was a high biological oxygen demand (BOD) and resulting low dissolved oxygen concentrations due to the decomposing submerged inundated rain forest and this was characteristic of conditions for the first seven years. For these first seven years the BOD in the epilimnion was relatively high and in the hypolimnion relatively low. After this time the BOD in the hypolimnion was increasing and in the epilimnion decreasing as a result of the accumulation of dead organic material in the hypolimnion (Richter and Nijssen, 1980).

However, the Aswan High Dam lake, which has a surface area of 5000 km^2, has become one of the main fisheries in Egypt and it is estimated that the annual catch could be as much as 100 000 tonnes, the main fish species being the tilapias and Nile perch (Abu-Zeid, 1983).

A reservoir may also reduce fluctuations in water levels. For example, a hydroelectric scheme on the Kafue River may destroy the floodplain environment which could seriously affect the numbers of wattled cranes (Douthwaite, 1974).

It can be seen that the environmental problems posed by water abstraction and hydroelectric schemes are both diverse and complex as well as unpredictable, even to the freshwater ecologist.

14 The future

We are capable of some remarkable technological feats. Just as this book is being finished the spacecraft Voyager 2 has swept past the planet Neptune, the last of its planetary encounters. Voyager's expedition has revealed bizarre worlds with their own weather systems. Some include water, though not in the life sustaining regime found on planet Earth. At the same time, in Russia, the Aral Sea, which is (or was) a very large lake some 64 500 km^2, has lost over 60 per cent of its water, since 1960, to a huge irrigation system. The result is a contracting shoreline that has left fishing villages over 100 km from the water, increasingly saline conditions and perhaps local climatic changes. All this despite the fact that the lake is listed by the International Biological Programme (IBP) and the International Union for the Conservation of Nature and Natural Resources (IUCN) as worthy of conservation on an international scale (Luther and Rzoska, 1971).

What we do with the Earth's water and aquatic habitats is a vital problem. Water, whether is be too little or too much, and the condition it is in, is a dominant fact of life and death for most of the world's population. The developed world has enjoyed a few illusory decades in the belief that plentiful, clean water was always available, but that notion is currently being rapidly eroded. It is customary for ecological texts to conclude with a bit of crystal ball gazing, maudlin optimism or doom and gloom and we will be no exception. However, rather than glib speculation about anything and everything the discussion of the future is concerned primarily with conservation at three levels: international, national and local.

14.1 International

The late twentieth century has seen an increasing internationalization of conservation efforts. There are three main reasons.

1. The problems themselves, such as acid rain or global climatic change due to greenhouse effects, are international in cause and effect. What

happens in one country may affect another country (such as acid deposition in Norway which was in large part derived from British emissions) or potentially every country.

2. The increasingly international organization of administrations, politics and the media has allowed trans-national discussion and policy. The European Economic Community has created a huge forum for environmental monitoring and the setting and enforcement of standards. On occasion these have met with resistance from national governments. The increasing appreciation of environmental problems and their international extent has been partly responsible for the growth of 'green' politics. This has ranged from activist organizations such as Friends of the Earth who operate around the world, to political parties in individual countries. The current public interest has resulted in a response by business with 'green' products available such as washing liquids and powders described as environmentally friendly by virtue of containing no phosphorus and being biodegradable. The 'green' consumer, identified by advertising agents as worthy of cultivation, may be restricted to the Western world but environmental issues have assumed prominence in the communist world. Happily, the Aral Sea has its own public campaign.

   On an even wider front in Europe, the Council of Europe, with its twenty-two member states, is actively involved in environmental conservation through its Environment, Conservation and Management Division and the Centre Naturopa based in Strasbourg. Much of this work is channelled through the Committee for the Conservation and Management of the Environment and Natural Habitats (CDPE). It is the watchdog for the efficient maintenance of National Parks and Reserves. Those achieving a standard of excellence receive the European Diploma. The committee also recommends specific projects, many of which concern aquatic habitats and their endangered flora and fauna. One of these is directed to the protection of dragonflies and their biotopes.

   The Council of Europe also takes a very keen interest in environmental education and organizes European teachers' seminars on specific environmental problems. One such seminar held in Sweden in 1986 was on the 'Acidification of Soil and Water' and was administered through the Council for Cultural Cooperation (CDCC) and the CDCC's Teacher Bursaries Scheme (Council of Europe, 1987).

3. Many conservation organizations working in the technical, practical sphere now operate internationally. Most are concerned with all habitats, for example World Wide Fund for Nature (WWF) and International Union for the Conservation of Nature. Some are specifically dealing with freshwaters notably the International Waterfowl and

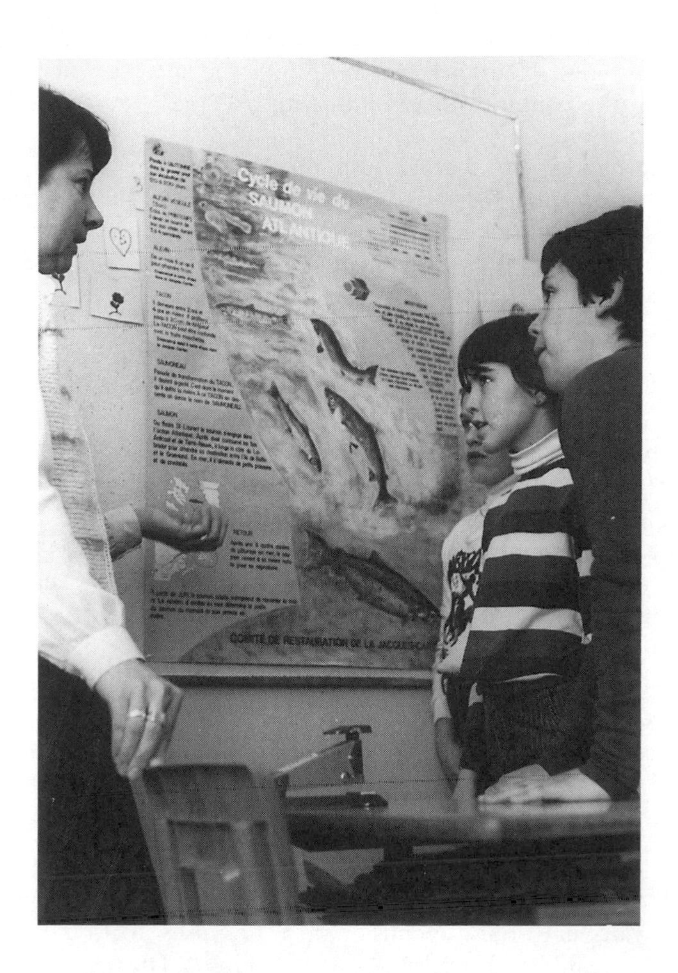

Plate 30 Conservation education should be an important part of the school syllabus. Here a teacher demonstrates the salmon life cycle to Quebec school children at a visitor centre set up by the Corporation of the Jacques-Cartier River. (Photo: M. Frenette)

Wetland Research Bureau (IWRB). This is closely linked to the Wildfowl Trust, recently renamed the Wildfowl and Wetlands Trust. The IWRB grew out of work to provide an international register of wetland sites and some protocol to conserve them. This resulted in the Ramsar convention, first signed in 1971 and some 40 countries are now members of the IWRB. A similar listing project, Project Aqua, was run by the IUCN and IBP, to describe lotic and lentic systems worthy of conservation at an international level. Such lists are useful, but as the Aral Sea suggests, no protection in themselves. The IWRB has recently established a Wetland Management Group to organize surveys, assessment and management of wetlands. The intention is not for the IWRB to do all the work, but to provide the necessary help and advice so that the countries themselves will take the responsibility for their conservation.

International-scale work also proceeds at the species level. The IUCN Red Data Books are source books on the status of endangered species, vertebrate and invertebrate. Many vertebrate species conservation projects exist. Invertebrate conservation relies primarily on habitat management.

At this international level the future for freshwater on this planet appears patchy. The organization and experience to maintain habitats exists and the technology to control extensive polluting emissions is available. The challenges are the will to do anything and who pays. In the Amazon, massive pollution from mercury residues derived from gold extraction has started with no sign of effective control. Even if developing countries are perfectly willing to run big conservation projects, or simply not develop tracts of land, the costs can be huge either as money spent or income not received. Conservation is increasingly tangled up with global economics.

Perhaps the situation is worthy of some optimism, if only because things are now being done, when, fifty years ago, there was nothing.

14.2 National

National problems and likely answers obviously vary with the nation in question. This brief discussion concentrates on the United Kingdom. The future of water in Britain depends on three premises. Firstly, there is a huge demand for water from industry and the public. There is an expectation that the water will be of a high quality. Secondly, industrial, agricultural and domestic sources of pollution are large but the facilities for monitoring and treatment are well developed. Thirdly, there is a well-established, expert suite of organizations concerned with water quality and conservation, many of essentially public status, for example Water Authorities in England and Wales, River Purification Boards in Scotland, and the Nature Conservancy Council (NCC). They have a good idea of the state of our water, the practices that maintain quality and good conservation management.

However, the Water Authorities have been split into commercial companies, dealing with the provision of water as a commercial business, and the newly created National Rivers Authority remaining as a public body responsible for pollution monitoring, control and conservation. Recently, the future of the NCC has also become uncertain. This national organization may be broken into separate regional bodies and merged with the

Countryside Commission.

So, the expertise and technology available nationally should be able to maintain water quality and conserve freshwater habitats but a combination of circumstances has contrived to give the British public a shock. The idea of clean, safe water in Britain is more recent than many people realize. In the nineteenth century many rivers, such as the Thames were a terrible source of disease. Industrial pollution remained conspicuous in some regions until only 20–30 years ago. Since then, clean water may have been generally available but circumstances have changed. Certain problems may have been predictable, such as the decay of once perfectly effective Victorian sewer systems and the steady build-up of agricultural nutrients in groundwater that now form a largely uncontrollable diffuse source of pollution. Other problems seem more fickle such as several recent drought summers and new disease threats associated with rats. The previously slow but steady impoverishment of Britain's rivers has halted with some declines evident. All this has coincided with the increased environmental awareness and the privatization of the industry.

Potentially, Britain is extremely well equipped to conserve its aquatic habitats. The Water Authorities had a national network of water quality monitoring and many regularly incorporated recent advances in sympathetic management techniques into their engineering. There are standard techniques for whole river corridor description (NCC, 1985). The NCC has increasingly been looking at water quality and wildlife, with a view to devising standards designed to maintain rivers of high conservation value and perhaps act as targets for restoration of other waters. Conservation organizations have national schemes for river and lake classification, representing clean water conditions and also perturbed habitats, such as the NCC's work with macrophytes and Freshwater Biological Associations' work with rivers (Wright *et al.*, 1984; Furse *et al.*, 1984). There is a national review of sites of particular conservation importance, akin to the international schemes (Ratcliffe, 1977). Many of these sites are in nature reserves or partially protected by registration as Sites of Special Scientific Interest. Outside of reserves we do now understand a great deal about the interactions of land use and freshwater and have expertise in good management and ameliorative practices (Solbé, 1986).

The national situation is similar to the international picture. There are increasing problems of water quantity and quality, but at the same time there is a very great deal of expertise and knowledge available. The problems are not what to do but finding the will to do it and pay for it. Unlike the international arena, where there is increasing cause for optimism, the immediate future in Britain may be bleak. The privatization of the water industry has raised the question of cost to a high profile. There may well be a period when high costs, due to work on old sewers and treatment plants plus new installations, coupled with public anxiety over prices squeeze money for conservation and management. At the same time the new National Rivers Authority and possible demise of the NCC, as a National organization, opens a lean period for effective conservation effort.

In Britain we can have high quality water and we can have a thriving aquatic wildlife, but only if we want it, work at it and pay for it.

14.3 Local

If the international and national scale of problems conform to the gloom school of environmentalism, the local scale offers a refreshing change.

Many problems such as pollution or major engineering, can only be effectively tackled by large organizations such as Water Authorities, but a great deal of conservation work can be done on a small scale. This level includes projects such as urban wildlife groups, wildlife trusts, conservation work parties, small-scale intervention by the Forestry Farming and Wildlife Advisory Group (FFWAG) and Ministry of Agriculture. Organizations not specifically concerned with the environment, such as schools and local councils, are heavily involved all the same, so too are individuals. The media also has a big part to play.

Plate 31 Creating a new aquatic habitat – pond construction underway (February 1987) in preparation for the Glasgow Garden Festival. (Photo: Scottish Wildlife Trust)

Plate 32 A new pond is born, complete with a diverse flora, in time for the 1988 Glasgow Garden Festival. (Photo: Scottish Wildlife Trust)

Even if the problems are beyond the practical powers of such groups, the effectiveness of campaigns and lobbying should not be underestimated and much of the power of many conservation bodies, such as The Royal Society for the Protection of Birds, stems from a vociferous membership. Suprisingly large-scale projects can be run. A potent mixture of pollution, children and the media has been conjured up in Britain to look at river pollution. In 1971 the Clean Stream Survey, using simplified biological monitoring kits issued to children, looked at rivers throughout the country, producing a basic but effective classification of quality (Mellanby, 1974). This project led to the formation of WATCH, an environmental club for children allied to the Wildlife and Naturalists Trusts. In 1987 the survey was repeated, spearheaded by the participation of a Sunday newspaper giving coverage in its colour magazine. Reports have been prepared and information packs devised for use in schools and other children's groups. From 1985 to 1987 WATCH has organized the Acid Drops project again with children nationwide testing the local rain for pH. The results are collated nationally. In 1987 children in Germany and Scandinavia joined in. The projects have attracted sponsorship from industry, for example British Telecom for WATCH on streams, British Petroleum for Acid Drops. The scientific value may be limited compared to complex studies and the sponsorship has been treated with cynicism but the mixture of action, education and publicity is a crucial part of conservation. If the main impression of national and international problems is that the future depends on our willingness to act and to pay, then the motivation fostered in children is vital.

Local action has also proven important for small-scale habitat conservation, restoration and creation. Sometimes this is largely accidental. Most garden ponds are primarily ornamental but have become important refuges for wildlife. Many organizations take an active role in lotic and lentic management. Expertise is widely available and several useful manuals exist (Lewis and Williams, 1984; Brooks and Agate, 1981). Small-scale conservation management is now encouraged in agriculture by FFWAG advisers. A recent reversal of previous Ministry of Agriculture policies, that seemed to ignore aquatic habitats, has led to the provision of useful advice leaflets to farmers on ponds and drainage. Urban environments have proved fruitful ground with urban wildlife groups in several cities and even inner-city reserves. Water has proved an especially attractive feature, with great benefits for education and recreation. Habitat can be linked to species. Amphibia have received increasing sympathetic attention. The result has been rescue of animals from ponds threatened with development and 'toad patrol', on roads that these animals have to cross to reach breeding sites, to save as many as possible from being run over. WATCH organizes a national Hopline, so that children can phone in with their sightings as the animals reappear in spring. Again a lot of help and advice is now available for work at this level (King and Clifford, 1985; Emery, 1986).

Plate 33 Helping to clean up our environment. Children and teaching staff from Holmlea Primary School, Cathcart, Glasgow, clear rubbish from the River Cart helped by parents. (Photo: *Glasgow Herald*)

It is easy to scoff at this level of activity. Hoplines and Acid Drops are not the stuff that eminent reports are made of, but they are the stuff that will encourage and enthuse the great majority of people who will have to pay for conservation. The spacecraft Voyager 2 is now heading out of our solar system. It has a plaque on its side so that should any space-aliens happen to find it they can navigate their way to Earth. As we hope this book shows, we do know a great deal about the water on this planet and the life it sustains; maybe not everything but enough so that we can manage the environment sensibly. Whether or not planet Earth will be in a fit state to impress visitors is not just a question of understanding our environment, it is also a question of having the determination to put these ideas into practice. Perhaps the most important message of this book is an optimistic one. There is a great deal you can do and we hope this short introduction provides encouragement.

References

Abo-Rady, M. D. K. (1980) Aquatic macrophytes as indicators for heavy metal pollution in the River Leine (West Germany). *Archives fur Hydrobiologie,* **89**, 387–404.

Abu-Zeid, M. (1983) The River Nile: Main water transfer projects in Egypt and impacts on Egyptian agriculture. In *Long-Distance Water Transfer. A Chinese Case Study and International Experiences* (eds. A. Biswas, Z. Dakang, J. Nickum and L. Changming). Tycoolgy International, Dun Laoghaire. pp. 16–34.

Alabaster, J. S. (ed.) (1985) *Habitat Modification and Freshwater Fisheries.* Butterworth, London.

Alabaster, J. S. and Lloyd, R. (1980) *Water Quality Criteria for Freshwater Fish.* Butterworth, London.

Allen, J. D. (1984) Hypothesis testing in ecological studies of aquatic studies. In *The Ecology of Aquatic Insects* (eds. V. H. Resh and D. M. Rosenberg). Praeger, New York.

Altshuller, A. P. and Linthurst, R. A. (eds.) (1985) The acidic deposition phenomenon and its effects: critical assessment review papers. Vol. II. Effects Science. US Environmental Protection Agency, Washington.

Angus, R. B. (1983) Evolutionary stability since the Pleistocene illustrated by reproductive compatibility between Swedish and Spanish *Helophorus lapponicus* Thornson (Coleoptera, Hydrophilidae). *Biological Journal of the Linnean Society,* **19**, 17–25.

Arber, A. A. (1920) *Water Plants.* Cambridge University Press, Cambridge.

Armitage, P. D. (1980) The effects of mine drainage and organic enrichment on benthos in the River Nent system, northern Pennines. *Hydrobiologia,* **74**, 119–28.

Atwell, R. (1970) Some effects of Lake Kariba on the ecology of a floodplain of the mid-Zambezi valley of Rhodesia. *Biological Conservation,* **2** (3).

Bacalbasa-Dobrovici, N. (1985) The effects on fisheries of non-biotic modifications of the environment in the east-Danube river area, In *Habitat Modification and Freshwater Fisheries* (ed. J. Alabaster). Butterworth, London, pp. 13–27.

Bailey-Watts, A. E. (1986) Seasonal variation in size spectra of phytoplankton assemblages in Loch Leven, Scotland. *Hydrobiologia,* **138**, 25–43.

Bailey-Watts, A. E. (1982) The composition and abundance of phytoplankton in Loch Leven (Scotland) 1977–1979 and a comparison with the succession in earlier years. *Internationale Revue der Gesamten Hydrobiologie Systematique Beihefte,* **67**, 1–25.

Balls, H., Moss, B. and Irvine, K. (1989) The loss of submerged plants with eutrophication I. Experimental design, water chemistry, aquatic plant and phytoplankton biomass in experiments carried out in ponds in the Norfolk Broads. *Freshwater Biology,* **22**, 71–87.

Bardach, J. E., Ryther, J. and McLarney, W. O. (1972) *Aquaculture.* John Wiley, New York, 868 pp.

Barica, J. and Mur, L. R. (eds.) (1980) *Hypertrophic Ecosystems. Developments in Hydrobiology,* Vol. 2. Junk, The Hague.

Barmuta, L. A. (1989) Habitat patchiness and macrobenthic community structure in an upland stream in temperate Victoria, Australia. *Freshwater Biology,* **21**, 223–36.

Barnes, L. E. (1983) The colonisation of Ball clay ponds by macroinvertebrates and macrophytes. *Freshwater Biology,* **13**, 561–78.

Baross, J. A., Dahm, C. N., Wards, A. K., Lilley, M. D. and Sedell, J. R. (1982) Initial microbiological response in lakes to Mt St Helens eruption. *Nature,* **296**, 49–52.

Barrera-Rodriguez, R., Machada-Allison, C. E., Bulla, L. and Strong, D. R. (1982) Mosquitoes and mourning in the Caracas cemetery. *Antenna,* **6**, 3.

Barth, E. (1988) *Reversibility of Acidification.* Elsevier, London.

Barton, B. A. (1977) Short-term effects of highway construction on the limnology of a small stream in southern Ontario. *Freshwater Biology,* **7**, 99–108.

Bates, A. J., Extence, C. A. and Forbes, W. J. (1985) *Biologically Based Water Quality Management.* Anglian Water, Lincoln Division.

Batten, L. A. (1977) Sailing on reservoirs and its effects on water birds. *Biological Conservation,* **11**, 49–58.

Batterbee, R. W. (1988) *Lake Acidification in the United Kingdom.* Ensis, London.

Batterbee, R. W., Flower, R. J., Stevenson, A. C. and Rippey, B. (1985) Lake acidification in Galloway: a palaeological test of competing hypotheses. *Nature*, **314**, 250–2.

Batterbee, R. W., Flower, R. J., Stevenson, A. C., Jones, V. J., Harriman, R. and Appleby, P. G. (1988) Diatom and chemical evidence for reversibility of acidification of Scottish lochs. *Nature*, **332**, 530–2.

Bayless, J. and Smith, W. B. (1967) The effects of channelization upon the fish population of lotic waters in eastern North Carolina. *Proceedings of the Annual Conference South-east Association Game and Fish Commission*, **18**, 230–8.

Bayly, I. A. E. and Williams, W. D. (1973) *Inland Waters and their Ecology*. Longman, Victoria, Australia.

Beadle, L. C. (1974) *The Inland Waters of Tropical Africa*. Longman. London.

Beaumont, P. (1975) Hydrology. In *River Ecology* (ed. B. A. Whitton). Blackwell, Oxford.

Beebee, T. J. C. (1983) Habitat selection by amphibians across an agricultural land–heathland transect in Britain. *Biological Conservation*, **27**, 111–24.

Begon, M., Harper, J. L. and Townsend, C. R. (1986) *Ecology: Individuals, Populations and Communities*. Blackwell, Oxford.

Bender, E. A., Case, T. J. and Gilpin, M. E. (1984) Perturbation experiments in community ecology: theory and practice. *Ecology*, **65**, 1–13.

Bennett, D. V. and Streams, F. A. (1986) Effects of vegetation on *Notonecta* (Hemiptera) distribution in ponds with and without fish. *Oikos*, **46**, 62–9.

Benson-Evans, K. and Williams, P. F. (1976) Transplanting aquatic bryophytes to assess river pollution. *Journal of Bryology*, **9**, 81–91.

Berglund, B. E. (ed.) (1986) *Handbook of Holocene Palaeoecology and Palaeohydrology*. Wiley, Chichester.

Beveridge, M. C. M. (1984) Cage and pen fish farming. Carrying capacity models and environmental impact. FAO Fisheries Technical Paper (255).

Bjork, S. (1972) Swedish lake restoration program gets results. *Ambio*, **1**, 153–65.

Bold, H. C. and Wynne, M. J. (1985) *Introduction to the Algae*. Prentice Hall, Englewood Cliffs.

Bronmark, C. (1985) Interactions between macrophytes, epiphytes and herbivores: an experimental approach. *Oikos*, **45**, 26–30.

Brook, A. J. (1964) The phytoplankton of the Scottish freshwater lochs. In *The Vegetation of Scotland* (ed. J. H. Burnett). Oliver and Boyd, Edinburgh.

Brookes, A. (1986) Response of aquatic vegetation to sedimentation downstream from river channelization works in England and Wales. *Biological Conservation,* **38**, 351–67.

Brookes, A. (1988) *Channelized Rivers. Perspectives for Environmental Management.* Wiley, Chichester and New York.

Brooks, A. and Agate, E. (1981) *Waterways and Wetlands. A Practical Conservation Handbook.* British Trust for Conservation Volunteers, Wallingford.

Brusven, M. A., Meehan, W. R. and Ward, J. F. (1986) Summer use of simulated banks by juvenile chinook salmon in an artificial Idaho channel. *North American Journal of Fisheries Management,* **6**, 32–7.

Burgis, M. J. and Morris, P. (1987) *The Natural History of Lakes.* Cambridge University Press, Cambridge.

Burton, M. A. S. and Peterson, P. J. (1979) Metal accumulation by aquatic bryophytes from polluted mine streams. *Environmental Pollution* (Series A), **19**, 39–46.

Callander, R. A. and Duder J. (1979) Reservoir sedimentation in the Rangitaiki River. *New Zealand Engineering,* **34**, 208–15.

Campbell, R. N. (1963) Some effects of impoundment on the environment and growth of brown trout (*Salmo trutta* L.) in Loch Garry, Inverness-shire. *Freshwater and Salmon Fisheries Research,* **30**.

Carpenter, K. E. (1928) *Life in Inland Waters with Especial Reference to Animals.* Sidgwick & Jackson, London, 267 pp.

Carpenter, S. R. and Lodge, D. M. (1986) Effects of submerged macrophytes on ecosystem processes. *Aquatic Botany,* **26**, 341–70.

Castenholz, R. W. and Wickstrom, C. E. (1975) Thermal streams. In *River Ecology* (ed. B. A. Whitton). Blackwell, Oxford.

Cave, J. D. (1985) The effects of the Kielder scheme on fisheries. *Journal of Fish Biology,* **27** (Suppl. A), 109–21.

Chang, W. Y. B. (1989) Integrated lake farming for fish and environmental management in large shallow Chinese lakes: a review. *Aquaculture and Fisheries Management,* **20**, 441–52.

Charter, E. (1988) *Survey of Spey Valley Lochs, 1985.* Nature Conservancy Contract Survey No. 16. Nature Conservancy Council, Peterborough.

Connell, J. H. (1980) Diversity and coevolution of competition, or the ghost of competition past. *Oikos*, **35**, 131–8.

Cooke, A. S. and Scorgie, H. R. A. (1983) *The Status of the Commoner Amphibians and Reptiles in Great Britain*. Focus on Nature Conservation No. 3. Nature Conservancy Council, Peterborough.

Coope, G. R. (1977) Fossil Coleopteran assemblages as sensitive indicators of climatic changes during the Devensian (last) Cold Stage. *Philosophical Transactions of the Royal Society, London, B,* **280**, 313–340.

Coope, G. R. (1986) Coleoptera analysis. In *Handbook of Holocene, Palaeoecology and Palaeohydrology* (ed. B. E. Berglund). Wiley, Chichester.

Corbet, P. S. (1982) *A Biology of Dragonflies*. Witherby, London.

Corkum, L. D. (1989) Patterns of benthic invertebrate assemblages in rivers of northwestern North America. *Freshwater Biology,* **21**, 191–206.

Council of Europe (1987) European Teachers' seminar on 'Acidification of soil and water – Field Studies'. Report DECS/EGT (86) 90–E. Council for Cultural Cooperation, Strasbourg.

Cowling, E. B. (1980) An historical resumé of progress in scientific and public understanding of acid precipitation and its biological consequences. Fragrapport 18/80, Sur Nedbors Virkning pa skog og Fisk: Norway.

Craig, J. F. and Kemper, J. B. (1987) *Regulated Streams. Advances in Ecology.* Plenum Press, New York.

Cresser, M. and Edwards, A. (1987) *Acidification of Freshwaters.* Cambridge University Press, Cambridge.

Crickmay, C. H. (1974) *The Work of the River.* MacMillan, London.

Crisp, D. T. (1984) Water temperature studies in the River North Tyne after impoundment by Kielder dam. Freshwater Biological Association, Teesdale Unit, Unpublished report.

Cryer, M. and Edwards, R. W. (1987) The impact of angler groundbait on benthic invertebrates and sediment respiration in a shallow eutrophic reservoir. *Environmental Pollution,* **46**, 137–150.

Cuker, B. E. (1983) Competition and co-existence among the grazing snail, *Lymnaea,* Chironomidae and microcrustacea in an Arctic epilithic lacustrine community. *Ecology,* **64**, 10–15.

Cummins, K. W. (1973) Trophic relations of aquatic insects. *Annual Review of Entomology,* **18**, 183–206.

Cummins, K. W. (1974) Structure and function of stream ecosystems. *Bioscience,* **24**, 631–41.

Currie, J. C. (1989) Tweed water quality. In *Tweed Towards 2000,* (ed. D. H. Mills). Tweed Foundation. Cambrian News, Aberystwyth.

Cyr. H. and Downing, J. A. (1988) The abundance of phytophilous invertebrates on different species of submerged macrophytes. *Freshwater Biology,* **20**, 365–74.

Dale, H. M. and Gillespie, T. J. (1977) The influence of submerged aquatic plants on temperature gradients in shallow water bodies. *Canadian Journal of Botany,* **55**, 2216–25.

Davies, C. (1988) Control and remedial strategies. In *Air Pollution, Acid Rain and the Environment* (ed. K. Mellanby). The Watt Committee on Energy. Elsevier, London.

De Dekker, P. and Williams, W. D. (eds.) (1986) *Limnology in Australia.* Longman, London.

Diamond, J. and Case, T. J. (1986) *Community Ecology.* Harper and Row, New York.

Dickson, M. (1988) Practical preventive and curative measures for aquatic ecosystems. In *Air Pollution and Ecosystems* (ed. P. Mathy). Reidel, Dordrecht.

Dillon, P. J. and Rigler, F. G. (1975) A simple method for predicting the capacity of a lake for development based on lake trophic status. *Journal of the Fisheries Research Board of Canada,* **32**, 1519–31.

Dillon, P. J., Reid, R. A. and Girard, R. (1986) Changes in the chemistry of lakes near Sudbury, Ontario, following reduction in SO_2 emissions. *Water, Air and Soil Pollution,* **31**, 59–65.

Douthwaite, R. J. (1974) An endangered population of wattled cranes (*Grus carunculatus*). *Biological Conservation,* **6**, 134–42.

Dvorak, J. and Best, E. P. H. (1982) Macroinvertebrate communities associated with the macrophytes of Lake Vechten: structural and functional relationships. *Hydrobiologia,* **95**, 115–26.

Ebel, W. J. (1985) Review of effects of environmental degradation on the freshwater stages of anadromous fish. In *Habitat Modification and Freshwater Fisheries* (ed. J. Alabaster). Butterworth, London, pp. 62–79.

Edwards, R. W. and Owen, M. (1962) The effects of plants on rivers IV. The oxygen balance of a chalk stream. *Journal of Ecology,* **50**, 207–20.

Elder, H. Y. (1966) Biological effects of water utilisation by hydro-electric schemes in relation to fisheries, with special reference to Scotland. *Proceedings of the Royal Society of Edinburgh,* **B, LXIX**, iii/iv, 246–71.

Elliott, J. M. (ed.) (1989) Wild brown trout: the scientific basis for their conservation and management. *Freshwater Biology,* **21**, 1–137.

Emery, M. (1986) *Promoting Nature in Cities and Towns.* Croom Helm, London.

Environmental Resources Limited (1983) *Acid Rain: A Review of the Phenomenon in the EEC and Europe.* Graham and Trotman, London.

Extence, C. A. and Ferguson, A. J. D. (1989) Aquatic invertebrate surveys as a water quality management tool in the Anglian Water region. *Regulated Rivers: Research & Management,* **4**, 139–46.

Extence, C. A., Bates, A. J., Forbes, W. J. and Barham, P. J. (1987) Biologically based water quality management. *Environmental Pollution,* **45**, 221–36.

Fairbairn, W. A. (1967) Erosion in the River Findhorn valley. *Scottish Geographical Magazine,* **83**, 46–52.

Fitz, R. B. (1968) Fish habitat and population changes resulting from impoundment of Clinch River by Melton Hill Dam. *Journal of the Tennessee Academy of Science,* **43**, 7–15.

Forestry Commission (1989) *Forests and Water: Guidelines.* Forestry Commission, Edinburgh.

Foster, P. L. (1982) Metal resistances of Chlorophyta from rivers polluted by heavy metals. *Freshwater Biology,* **12**, 41–61.

Frank, J. H. and Lounibus, L. P. (1983) *Phytotelmata.* Plexus, New York.

Furse, M. T., Moss, D., Wright, J. F. and Armitage, P. D. (1984) The influence of seasonal and taxonomic factors on the ordination and classification of running-water sites in Great Britain and on the prediction of their macro-invertebrate communities. *Freshwater Biology,* **14**, 257–80.

Gee, J. H. R. and Giller, P. S. (1987) *Organization of Communities Past and Present.* Blackwell, Oxford.

George, D. G. and Edwards, R. W. (1974) Population dynamics and production of *Daphnia hyalina* in a eutrophic reservoir. *Freshwater Biology,* **4**, 445–66.

Gill, J. G. S. (1989) The changing scene of British forestry. *Sylva,* Centenary Edition, Department of Forestry & Natural Resources, University of Edinburgh, 11–14.

Gliwicz, Z. (1986) Predation and the evolution of vertical migration in zooplankton. *Nature,* **320**, 746–8.

Gonzales, E. A. and Vaidya, B. S. (1963) On the larvicidal properties of charophytes. *Hydrobiologia,* **21**, 188–92.

Gorman, O. T. and Karr, J. R. (1978) Habitat structure and stream fish communities. *Ecology,* **59**, 502–15.

Gosling, L. M., Baker, S. J. and Clarke, C. N. (1988) An attempt to remove coypus (*Myocastor coypus*) from a wetland habitat in East Anglia. *Journal of applied Ecology,* **25**, 49–62.

Greene, G. E. (1950) Land use and trout streams. *Journal of Soil and Water Conservation,* **5**, 125–6.

Gregg, W. W. and Rose, F. L. (1982) The effects of aquatic macrophytes on the stream microenvironment. *Aquatic Botany,* **14**, 309–24.

Grimås, U. (1962) The effect of increased water level fluctuations upon the bottom fauna in Lake Blåsjön, Northern Sweden. *Report of the Institute of Freshwater Research, Drottningholm,* **44**, 14–41.

Hall, J. D. and Lantz, R. L. (1969) Effects of logging on the habitat of coho salmon and cutthroat trout in coastal streams. In *Symposium on Salmon and Trout in Streams* (ed. T. G. Northcote). H. R. MacMillan Lectures in Fisheries, 1968, University of British Columbia, Vancouver, pp. 355–75.

Halstead, B. G. and Tash, J. C. (1982) Unusual diel pHs in water as related to aquatic vegetation. *Hydrobiologia,* **96**, 217–24.

Ham, S. F., Cooling, D. A., Hiley, P. D., McLeish, P. R., Scorgie, H. R. A. and Berrie, A. D. (1982) Growth and recession of aquatic macrophytes on a shaded section of the River Lambourn, England, from 1971 to 1980. *Freshwater Biology,* **12**, 1–15.

Hammerton, D. (1988) Freshwater. In *Air Pollution, Acid Rain and the Environment* (ed. K. Mellanby). The Watt Committee on Energy. Elsevier, London.

Harriman, R. (1978) Nutrient leaching from fertilised watersheds in Scotland. *Journal of applied Ecology,* **15**, 933–42.

Harriman, R. and Wells, D. E. (1985) Causes and effects of surface acidification in Scotland. *Journal of Water Pollution Control,* **84**, 215–22.

Hart, D. D. (1986a) The adaptive significance of territoriality in filter feeding larval blackflies (Diptera: Simulidae). *Oikos,* **46**, 88–92.

Hart, D. D. (1986b) Do experimental studies of patch use provide evidence of competition in stream insects? *Oikos,* **47**, 123–125.

Harvey, I. F. and Corbet, P. S. (1986) Territorial interactions between larvae of dragonfly *Pyrrhosoma nymphula:* outcome of encounter. *Animal Behaviour,* **34**, 1550–61.

Haslam, S. M. (1978) *River Plants.* Cambridge University Press, Cambridge.

Haslam, S. M. and Wolseley, P. (1981) *River Vegetation; its Identification, Assessment and Management.* Cambridge University Press, Cambridge.

Heads, P. A. (1985) The effect of invertebrate predation on the foraging movements of *Ischnuria elegans* larvae (Odonata: Zygoptera). *Freshwater Biology,* **15**, 559–71.

Healy, M. (1984) Fish predation on aquatic insects. In *The Ecology of Aquatic Insects* (eds. V. H. Resh and D. M. Rosenberg). Praeger, New York.

Hellawell, J. M. (1978) *Biological Surveillance of Rivers. A Biological Monitoring Handbook.* Water Research Centre, Stevenage.

Hellawell, J. M. (1986) *Biological Indicators of Freshwater Pollution and Environmental Management.* Elsevier, London.

Henriques, P. R. (ed.) (1987) *Aquatic Biology and Hydro-electric Power Development in New Zealand.* Oxford University Press, Auckland.

Hepher, B. and Pruginin, Y. (1981) *Commercial Fish Farming with Special Reference to Fish Culture in Israel.* John Wiley, New York.

Hickling, C. F. (1971) *Fish Culture* (revised edition). Faber, London.

Hildrew, A. G. and Townsend, C. R. (1980) Aggregation, interference and foraging by larvae of *Plectronema conspersa* (Trichoptera: Polycentropidae). *Animal Behaviour,* **28**, 553–60.

Hildrew, A. G. and Townsend, C. R. (1987) Organization in freshwater benthic communities. In *Organization of Communities Past and Present* (eds. J. H. R. Gee and P. S. Giller). Blackwell, Oxford.

Hildrew, A. G., Townsend, C. R., Francis, J. and Finch, K. (1984) Cellulytic decomposition in streams of contrasting pH and its relationship with invertebrate community structure. *Freshwater Biology,* **14**, 323–8.

Holdgate, M. W. (1979) *A Perspective of Environmental Pollution.* Cambridge University Press, Cambridge.

Holmes, N. T. H. (1983) *Typing British Rivers According to their Flora.* Focus on Nature Conservation No. 4. Nature Conservancy Council, Peterborough.

Holmes, N. T. H. and Newbold, C. (1984) *River Plant Communities – Reflectors of Water and Substrate Chemistry.* Report No. 9. Nature Conservancy Council, Peterborough.

Hornung, M. (1988) Effects of land management on acidification of aquatic ecosystems and the implications for the development of ameliorative measures. In *Air Pollution and Ecosystems* (ed. P. Mathy). Reidel, Dordrecht.

Howells, G. (ed.) (1986) Loch Fleet Report. A report of the pre-intervention phase 1984–86. CEGB, SSEB, NSHEB and British Coal.

Howells, W. R. and Merriman, R. (1986) Pollution from agriculture in the area of the Welsh Water Authority. In *Effects of Land Use on Fresh Waters* (ed. J. F. Solbé). Ellis Horwood, Chichester, pp. 267–82.

Huet, M. (1954) Biologie, profils en long et en travers des eaux courantes. *Bulletin Francais Pisciculture,* **175**, 41–53.

Huet, M (1986) *Textbook of Fish Culture.* Fishing News Books, Farnham.

Hughey, K. F. D. (1987) Wetland birds, in *Aquatic Biology and Hydro-electric Power Development in New Zealand* (ed. P. R. Henriques), 264–275, Oxford University Press, Auckland.

Hunter, M. L. jr, Jody, J. J. and Moring, J. R. (1986) Duckling responses to lake acidification. Do black ducks and fish compete? *Oikos,* **47**, 26–32.

Hutchinson, G. E. (1975) *A Treatise on Limnology,* Vols, I, II and III. Wiley, New York.

Hynes, H. B. N. (1970) *The Ecology of Running Waters.* University of Liverpool Press, Liverpool.

Irvine, K., Moss, B. and Balls, H. (1989) The loss of submerged plants with eutrophication II. Relationships between fish and zooplankton in a set of experimental ponds, and conclusions. *Freshwater Biology,* **22**, 88–108.

Irvine, J. R. and Jowett, I. G. (1987) Flow control, in *Aquatic Biology and Hydro-electric Power Development in New Zealand* (ed. P. R. Henriques), 94–112. Oxford University Press, Auckland.

James, A. and Evison, L. (1979) *Biological Indicators of Water Quality.* Wiley, Chichester.

Jeffries, M. J. (1989) Measuring Talling's 'element of chance in pond populations'. *Freshwater Biology,* **21**, 383–93.

Jeffries, M. J. (1990a) Interspecific differences in movement and hunting success in damselfly larvae (Zygoptera: Insecta): responses to prey availability and predation threat. *Freshwater Biology,* **23**, 191–96.

Jeffries, M. J. (1990b) Evidence of induced plant defences in a pondweed. *Freshwater Biology* (in press).

Jeffries, M. J. and Lawton, J. H. (1984) Enemy-free space and the structure of ecological communities. *Journal of the Linnaen Society*, **23**, 269–86.

Jenkins, R. A., Wade, K. R. and Pugh, E. (1984) Macroinvertebrate habitat relationships in the River Teifi catchment and the significance to conservation, *Freshwater Biology*, **14**, 23–42.

Jinghua, W. and Yongke, L. (1983) An investigation of the water quality and pollution in the rivers of the proposed water transfer region, In *Long-Distance Water Transfer: A Chinese Case Study and International Experiences* (eds. A. Biswas, Z. Dakang, J. Nickum and L. Changming) Tycooly International, Dun Laoghaire. pp. 362–71.

Johnson, R. K. and Wiederholm, T. (1989) Classification and ordination of profundal macroinvertebrate communities in nutrient poor, oligo-meshamic lakes in relation to environmental data. *Freshwater Biology*, **21**, 375–86.

Jonasson, M. P. (ed.) (1979) Ecology of eutrophic, subarctic Lake Myvatn and the River Laxa. *Oikos*, **32**, 1–2.

Jorgensen, S. E. (1980) *Lake Management, Water Development, Supply and Management, Developments in Hydrobiology*, Vol. 14. Pergamon, Oxford.

Jowett, I. G. (1983) Siltation of the lower Waitaki riverbed. Power Division, Ministry of Works and Development.

Jowett, I. G. and Hicks, D. M. (1981) Surface suspended and bedload sediment, Clutha River System. *Journal of Hydrology (NZ)*, **20** (2), 121–130.

Jupp, B. P. and Spence, D. H. N. (1977) Limitations on macrophytes in a eutrophic lake, Loch Leven, I. Effects of phytoplankton. *Journal of Ecology*, **65**, 175–86.

Kalliola, R. and Puhakka, M. (1988) River dynamics and vegetation mosaicism: a case study of the River Kamajohka, northernmost Finland. *Journal of Biogeography*, **15**, 703–19.

Kassas, M. (1974) The River Nile Ecological System: A study towards an international programme. *Biological Conservation*, **4** (1).

Kaatra, K. and Simola, D. (1985) Water level regulation of Lake Inari: impacts and compensation measures. In *Habitat Modification and Freshwater Fisheries* (ed. J. Alabaster). Butterworth, London, pp. 173–8.

Kennedy, G. J. A. (1987) Silage effluent pollution – costs and prevention. *Agriculture in Northern Ireland*, **60** (12), 5 pp.

Kerfoot, W. C. and Sih, A. (1987) *Predation. Direct and Indirect Impacts on Aquatic Communities.* University Press of New England, Hanover.

Kew, H. W. (1983) *The Dispersal of Shells,* London.

King, A. and Clifford, S. (1985) *Holding Your Own. An action guide to local conservation.* Maurice Temple Smith, London.

Kirk, J. T. O. (1983) *Light and Photosynthesis in Aquatic Ecosystems.* Cambridge University Press, Cambridge.

Krenkel, P. A. and Parker, L. F. (1969) *Biological Aspects of Thermal Pollution.* Vanderbilt University Press, Cambridge.

Lamberti, G. A. and Resh, V. H. (1983) Stream periphyton and insect herbivores: an experimental study of a caddisfly population. *Ecology,* **64**, 1124–35.

Lamberti, G. A. and Moore, J. W. (1984) Aquatic insects as primary consumers. In *The Ecology of Aquatic Insects* (eds. V. H. Resh and D. M. Rosenberg). Praeger, New York.

Langford, T. E. (1983) *Electricity Generation and the Ecology of Natural Waters.* Liverpool University Press, Liverpool.

Lever, C. (1979) *The Naturalised Animals of the British Isles.* Granada, London.

Lewin, J. (ed.) (1981) *British Rivers.* Allen and Unwin, London.

Lewis, G. and Williams, G. (1984) *Rivers and Wildlife Handbook: A Guide to Practices which Further the Conservation of Wildlife in Rivers.* Royal Society for the Protection of Birds and Royal Society for Nature Conservation, Sandy, Bedfordshire.

Liddle, M. J. and Scorgie, H. R. A. (1980) The effects of recreation on freshwater plants and animals: A review. *Biological Conservation,* **17**, 183–206.

Light, J. J. (1975) Clear lakes and aquatic bryophytes in the mountains of Scotland. *Journal of Ecology,* **63**, 937–43.

Lindstrom, T. (1973) Life in a lake reservoir. *Ambio,* (**2**) 5, 145, 148–53.

Livermore, D. F. and Wunderlich, W. E. (1969) Mechanical removal of organic production from waterways. In *Eutrophication: Causes, Consequences and Correctives* (ed. G. A. Rohlich). National Academy of Sciences, Washington.

Lodge, D. M. (1985) Macrophyte–gastropod associations: observations and experiments on macrophyte choice by gastropods. *Freshwater Biology,* **15**, 695–708.

Lowe-McConnell, R. (1966) *Man-Made Lakes.* Academic Press, London.

Lowe-McConnell, R. (1975) *Fish Communities in Tropical Freshwaters.* Longman, London.

Luther, H. and Rzoska, J. (1971) *Project Aqua: a source book of inland waters proposed for conservation.* International Biological Programme, Blackwell, Oxford.

Macan, T. T. (1977) Changes in the vegetation of a moorland fish pond in twenty-one years. *Journal of Ecology,* **65**, 95–106.

MacArthur, R. H. and Wilson, E. O. (1967) *The Theory of Island Biogeography.* Princeton University, Princeton.

McAuliffe, J. R. (1984) Resource depression by a stream herbivore: effects on distribution and abundances of other grazers. *Oikos,* **42**, 327–33.

McCahon, C. P., Carling, P. A. and Pascoe, D. (1987) Chemical and ecological effects of a peat slide: *Environmenal Pollution*, **45**, 275–89.

McCarthy, D. T. (1983) The impact of arterial drainage on fish stocks in the Trimblestown River. *Advances in Fish Biology in Ireland* (ed. C. Moriarty). Irish Fisheries Investigations, Series A, No. 23, pp. 16–19.

McCarthy, D. T. (1985) The adverse effects of channelisation and their amelioration. In *Habitat Modification and Freshwater Fisheries* (ed. J. Alabaster). Butterworth, London, pp. 83–97.

MacDonald, S. M. and Mason, C. F. (1983) Some factors influencing the distribution of otters (*Lutra lutra*). *Mammal Review,* **13**, 1–10.

McGrath, C. J. and Murphy, D. F. (1965) Engineering investigations into the effects of the harnessing of the River Lee, Co. Cork, Ireland, for hydro-electric purposes on the habitat and migration of salmonid stocks in that river system. *International Council for the Exploration of the Sea,* C.M. 1965/M:41.

McVean, D. N. and Lockie, J. D. (1969) *Ecology and Land Use in Upland Scotland.* University Press, Edinburgh.

Maitland, P. S. and Turner, A. K. (1987) *Angling and Wildlife in Fresh Waters.* Institute of Terrestrial Ecology, Grange over Sands.

Maitland, P. S., Lyle, A. A. and Campbell, R. N. B. (1987) *Acidification and Fish in Scottish Waters.* Institute of Terrestrial Ecology, Grange over Sands.

Marchant, J. H. and Hyde, P. M. (1979) Population changes for waterways birds. *Bird Study,* **26**, 227–39.

Marchant, J. H. and Hyde, P. M. (1980) Aspects of the distribution of riparian birds in Britain and Ireland. *Bird Study.* **27**, 183–202.

Mason, C. F. and Bryant, R. J. (1975) Changes in the ecology of the Norfolk Broads. *Freshwater Biology,* **5**, 257–70.

Mason, C. F. and MacDonald, S. M. (1982) The input of terrestrial invertebrates from tree canopies to a stream. *Biological Conservation,* **12**, 305–11.

Mason, C. F. and MacDonald, S. M. (1986) *Otters; Ecology and Conservation.* Cambridge University Press, Cambridge.

Mason, C. F., MacDonald, S. M. and Hussey, A. (1984) Structure, management and conservation value of the riparian woody plant community. *Biological Conservation,* **29**, 201–16.

Melack, J. M. (ed.) (1988) Saline lakes. *Hydrobiologia,* **158**, 1–316.

Mellanby, K. (1974) A water pollution survey, mainly by British schoolchildren. *Environmental Pollution,* **6**, 161.

Mellanby, K. (ed.) (1988) *Air Pollution, Acid Rain and the Environment.* The Watt Committee on Energy Report No. 18. Elsevier, London.

Merican, Z. O. and Phillips, M. J. (1985) Solid waste production from rainbow trout *Salmo gairdneri* Richardson, cage culture. *Aquaculture and Fisheries Management,* **16**, 55–70.

Merry, E. R. (1985) Pollution from farms. Atlantic Salmon Trust, Progress Report, September, p. 25.

Meynell, P. J. (1973) A hydrobiological survey of a small Spanish river grossly polluted by oil refinery and petrochemical wastes. *Freshwater Biology,* **3**, 503–520.

Miller, J. C. (1986) Manipulations and interpretations in tests, for competition in streams: "controlled" vs "natural" experiments. *Oikos,* **47**, 120–123.

Mills, D. H. (1964) The ecology of the young stages of the Atlantic salmon in the River Bran, Ross-shire. *Freshwater and Salmon Fisheries Research, Scotland,* **32**.

Mills, D. H. (1969) The survival of juvenile Atlantic salmon and brown trout in some Scottish streams. In *Symposium on Salmon and Trout in Streams.* (ed. T. G. Northcote). H. R. MacMillan Lectures in Fisheries, 1968, University of British Columbia, Vancouver, pp. 217–28.

Mills, D. H. (1971) *Salmon and Trout: A Resource, its Ecology, Conservation and Management.* Oliver and Boyd, Edinburgh.

Mills, D. H. (1989a) *Ecology and Management of Atlantic Salmon.* Chapman and Hall, London.

Mills, D. H. (1989b) The fish populations of Tweed – their distribution and interaction in a changing environment, In *Tweed Towards 2000* (ed. D. H. Mills). Tweed Foundation. Cambrian News, Aberystwyth.

Mills, D. H. (1990) The interaction between aquaculture and wild fisheries. 20th Annual Study Course of the Institute of Fisheries Management, Regional Technical College, Galway, Republic of Ireland.

Milner, N. J., Hemsworth, R. J. and Jones, B. E. (1985) Habitat evaluation as a fisheries management tool. *Journal of Fish Biology,* **27** (Suppl. A), 85–108.

Minshall, G. W. (1984) Aquatic insect–substratum relationships, In *The Ecology of Aquatic insects* (eds. V. H. Resh and D. M. Rosenberg). Praeger, New York.

Minshall, G. W., Andrews, D. A. and Manuel-Faler, C. Y. (1983a) Application of island biogeographic theory to streams: macroinvertebrate recolonization of the Teton River, Idaho. In *Stream Ecology* (eds. J. R. Barnes, and G. W. Minshall). Plenum, New York.

Minshall, G. W., Petersen, R. C., Cummins, K. W., Bott, T. L., Sedell, J. R., Cushing, C. E. and Vannote, R. L. (1983b) Interbiome comparison of stream ecosystem dynamics. *Ecological Monographs,* **53**, 1–25.

Moore, J. W. and Ramamoorthy, S. (1984) *Heavy Metals in Natural Waters.* Springer-Verlag, New York.

Morisawa, M. (1985) *Rivers, Form and Process.* Longman, London.

Moss, B. (1977) Conservation problems in the Norfolk Broads and rivers of East Anglia, England – phytoplankton, boats and the causes of turbidity. *Biological Conservation,* **12**, 95–114.

Moss, B. (1980a) Further studies on the palaeolimnology and changes in the phosphorus budget of Barton Broad, Norfolk. *Freshwater Biology,* **10**, 261–79.

Moss, B. (1980b) *Ecology of Fresh Waters.* Blackwell, Oxford.

Moss, B. (1983) The Norfolk Broads experiments in the restoration of a complex wetland. *Biological Review,* **58**, 521–61.

Moss, B. and Leah, R. T. (1982) Changes in the ecosystem of a guanotrophic and brackish shallow lake in eastern England: potential problems in its restoration. *Internationale Revue der gesamten Hydrobiologie,* **67**, 625–59.

Moss, B., Forrest, D. E. and Phillips, G. (1979) Eutrophication and palaeolimnology of two small mediaeval man-made lakes. *Archives fur Hydrobiologie,* **85**, 409–25.

Moss, B., Halls, H., Irvine, K. and Stansfeld, J. (1986) Restoration of two lowland lakes by isolation from nutrient-rich water sources with and without removal of sediment. *Journal of applied Ecology,* **23**, 391–414.

Moss, D., Furse, M. T., Wright J. F. and Armitage, P. D. (1987) The prediction of the macro-invertebrate fauna of unpolluted running water sites in Great Britain using environmental data. *Freshwater Biology,* **17**, 41–52.

Muirhead-Thomson, R. C. (1987) *Pesticide Impact on Stream Fauna*. Cambridge University Press, Cambridge.

Munawar, M. and Talling, J. F. (eds.) (1986) The seasonality of freshwater phytoplankton: a global perspective. *Hydrobiologia,* **138**, 1–221.

Murdoch, W. W., Scott, M. A. and Ebsworth, P. (1984) Effects of the general predator *Notonecta* (Hemiptera) upon a freshwater community. *Journal of Animal Ecology,* **53**, 791–808.

Murphy, K. J. and Eaton, J. W. (1983) The effect of pleasure boat traffic on macrophyte growth in canals. *Journal of applied Ecology,* **20**, 713–29.

Murphy, K. J., Hanbury, R. G. and Eaton, J. W. (1981) The ecological effects of 2-methylthiotriazine herbicides used for aquatic weed control in navigable canals. I. Effects on aquatic flora and water chemistry. *Archives fur Hydrobiologie,* **91**, 294–331.

Nature Conservancy Council, (1985) *Surveys of Wildlife in River Corridors: Draft Methodology*. NCC, Peterborough.

Naumann, E. (1919) Nagra synpunkter angaende planktons okolgie. Med. sraskild hansyn till fytoplankton. *Svensk Botanisk Tidskrift,* **13**, 129–58.

Newbold, C. (1975) Herbicides in aquatic systems. *Biological Conservation,* **7**, 97–118.

Newbold, C., Purseglove, J. and Holmes, N. (1983) *Nature conservation and river engineering*. Nature Conservancy Council: Peterborough.

Nygaard, G. (1949) Hydrobiological studies of some Danish ponds and lakes. II. The Quotient Hypothesis, and some new or little known phytoplankton organisms. K. danske vidensk Selsk. **8**.

Nystrom, U. and Hultberg, H. (1988) Effects of hydrology on the reacidification of the limed lake Gardsjon. In *Air Pollution and Ecosystems* (ed. P. Mathy). Reidel, Dordrecht.

Ormerod, S. J. (1988) The microdistribution of aquatic invertebrates in the Wye river system: the result of abiotic or biotic factors? *Freshwater Biology,* **20**, 241–7.

Ormerod, S. J., Allinson, N., Hudson, D. and Tyler, S. J. (1986) The distribution of dippers (*Cinclus cinclus* (L.): Aves) in relation to stream acidity in upland Wales. *Freshwater Biology,* **16**, 501–8.

Ormerod, S. J., Wade, K. R. and Gee, A. S. (1987) Macro-floral assemblages in upland Welsh streams in relation to acidity and their importance to invertebrates. *Freshwater Biology,* **18**, 545–57.

Ostrofsky, M. L. and Zettler, E. R. (1986) Chemical defences in aquatic plants. *Journal of Ecology*, **74**, 279–87.

Paine, R. T. (1966) Food web complexity and species diversity. *American Naturalist,* **100**, 65–75.

Palmer, M. (1989) *A Botanical Classification of Standing Waters in Great Britain.* Nature Conservancy Council Surveys in Nature Conservation No. 19. NCC, Peterborough.

Peckarsky, B. L. (1984) Predator–prey interactions among aquatic insects. In *The Ecology of Aquatic Insects* (eds. V. H. Resh and D. M. Rosenberg). Praeger, New York.

Peckarsky, B. L. and Dodson, S. I. (1980) Do stonefly predators influence benthic distributions in streams? *Ecology,* **61**, 1275–82.

Peterson, R. C. and Cummins, K. W. (1974) Leaf processing in a woodland stream. *Freshwater Biology,* **4**, 343–68.

Phillips, G. L., Eminson, D. F. and Moss, B. (1978) A mechanism to account for macrophyte decline in progressively eutrophicated freshwaters. *Aquatic Botany,* **4**, 103–26.

Phillips, M. (1984) Environmental effects of fish farming: implications for wild salmonids. *Proceedings of the 15th Annual Study Course of the Institute of Fisheries Management,* University of Stirling.

Phillips, M. J. (1985) The environmental impact of cage culture on Scottish freshwater lochs. Institute of Aquaculture, University of Stirling.

Phillips, M. J. and Beveridge, M. C. M. (1986) Cages and the effect on water conditions. *Fish Farmer*, **9**, 17–19.

Phillips, M. J., Beveridge, M. C. M. and Ross, L. G. (1985) The environmental impact of salmonid cage culture on inland fisheries: present status and future trends. *Journal of Fish Biology,* **27** (Suppl. A), 123–37.

Poff, N. L. and Matthews, R. A. (1986) Benthic macroinvertebrate community structural and functional group response to thermal enhancement in the Savannah River and a coastal plain tributary. *Archives fur Hydrobiologie,* **106**, 119–37.

Pond Action (1988) Pond Action News No. 1. Oxford Polytechnic, Oxford.

Rask, M., Heinanen, A., Salonene, K., Arrola, L., Bergstrom, I., Liukkonen, M. and Ojala, A. (1986) The limnology of a small, naturally acidic highly humic forest lake. *Archives fur Hydrobiologie,* **196**, 351–71.

Ratcliffe, D. A. (ed.) (1977) *A Nature Conservation Review.* Cambridge University Press, Cambridge.

Ratcliffe, D. A. and Oswald, P. H. (eds.) (1988) *The Flow Country.* Nature Conservancy Council, Peterborough.

Raven, J. A. (1984) *Energetics and Transport in Aquatic Plants.* Alan Liss, New York.

Raven, P. J. (1986a) Changes in waterside vegetation following two-stage channel construction on a small rural clay river. *Journal of applied Ecology,* **23**, 989–1000.

Raven, P. J. (1986b) Vegetation changes within the flood relief stage of two-stage channels excavated along a small rural clay river. *Journal of applied Ecology,* **23**, 1001–12.

Reynolds, C. S. (1987) Community organization in the freshwater plankton. In *Organization of Communities Past and Present* (eds. J. H. R. Gee and P. S. Giller). Blackwell, Oxford.

Richards, K. (1982) *Rivers, Form and Process in Alluvial Channels.* Methuen, London.

Richter, C. J. J. and Nijssen, H. (1980) Notes on the fishery potential and fish fauna of the Brokopondo Reservoir (Surinam). *Fisheries Management,* **11** (3). 119–30)

Riessen, H. P. (1984) The other side of cyclomorphosis: why *Daphnia* lose their helmets. *Limnology and Oceanography,* **29**, 1123–7.

Rooke, J. B. (1984) The invertebrate fauna of four macrophytes in a lotic system. *Freshwater Biology,* **14**, 507–13.

Room, P. M., Harley, K. L. S., Forno, I. W. and Sands, D. P. A. (1981) Successful biological control of the floating weed *salvinia. Nature,* **294**, 78–80.

Round, F. E. (1981) *The Ecology of Algae.* Cambridge University Press, Cambridge.

Round, P. D. and Moss, M. (1984) The waterbird populations in three Welsh rivers. *Bird Study,* **31**, 61–8.

Sand-Jensen, K. and Madsen, T.V. (1989) Invertebrates graze submerged rooted macrophytes in lowland streams. *Oikos,* **55**, 420–3.

Scorgie, H. R. A. (1980) Ecological effects of the aquatic herbicide Cyanatryn on a drainage channel. *Journal of applied Ecology,* **17**, 207–25.

Sculthorpe, C. D. (1967) *The Biology of Aquatic Vascular Plants.* Arnold, London.

Shapiro, J. and Wright, D. I. (1984) Lake restoration by biomanipulation: Round Lake, Minnesota, the first two years. *Freshwater Biology,* **14**, 371–83.

Sheldon, A. L. (1984) Colonization dynamics of aquatic insects. In *The Ecology of Aquatic Insects* (eds. V. H. Resh and D. M. Rosenberg). Praeger, New York.

Singer, R., Roberts, D. A. and Baylen, C. W. (1983) The macrophyte community of an acidic lake in Adirondack (New York, USA): a new depth record for aquatic angiosperms. *Aquatic Botany,* **16**, 49–57.

Sissons, J. B. (1976) *Scotland. The Geomorphology of the British Isles.* Methuen, London.

Sladecek, V. (1979) Continental systems for the assessment of river water quality. In *Biological Indicators of Water Quality.* (eds. A. James and L. Evison). Wiley, London.

Smith, B. D. (1980) The effects of afforestation on the trout of a small stream in southern Scotland. *Fisheries Management,* **11**, 39–58.

Smith, B. D., Cuttle, S. P. and Maitland, P. S. (1981) The profundal zoobenthos, In *The Ecology of Scotland's Largest Lochs Lomond, Awe, Ness, Morar and Shiel* (ed. P. S. Maitland). Junk, The Hague, 297 pp.

Smith, I. and Lyle, A. (1979) *Distribution of Fresh Waters in Great Britain.* Institute of Terrestrial Ecology, Edinburgh.

Solbé J. F. (1982) Fish farm effluents – cause for concern? *Water,* March 1982, 22–5.

Solbé, J. F. (ed.) (1986) *Effects of Land Use on Freshwaters.* Water Research Centre. Ellis, Hornwood, Chichester.

Solbé, J. F. (1987) European Inland Fisheries Advisory Commission working party on fish-farm effluents, *Water Research Centre Environment,* Medmenham.

Solbé, J. F. (1988) *Water Quality for Salmon and Trout,* Altantic Salmon Trust, Pitlochry.

Spence, D. H. N. (1964) The macrophytic vegetation of freshwater lochs, swamps and associated fens. In *The Vegetation of Scotland* (ed. J. H. Burnett). Oliver and Boyd, Edinburgh.

Spillett, P. B., Armstrong, G. S. and Magrath, P. A. G. (1985) Ameliorative methods to reinstate fisheries following land drainage operations, In *Habitat Modification and Freshwater Fisheries* (ed. J. Alabaster). Butterworth, London. pp. 124–30.

Stanford, J. A. and Ward, J. V. (1983) Insect species diversity as a function of environmental variability and disturbance in stream systems. In *Stream Ecology* (eds. J. R. Barnes and G. W. Minshall). Plenum, New York.

Stansfield, J., Moss, B. and Irvine, K. (1989). The loss of submerged plants with eutrophication III. Potential role of organochlorine pesticides: a palaeoecological study. *Freshwater Biology,* **22**, 109–132.

Stewart, L. (1963) *Investigations into Migratory Fish Propagation in the Area of the Lancashire River Board.* Barber, Lancaster, 80 pp.

Stich, H-B. and Lampert, W. (1981) Predator evasion as an explanation of diurnal vertical migration by zooplankton. *Nature*, **293**, 396–8.

Stockner, J. G. (1967) Observations of thermophilic algal communities in Mount Rainier and Yellowstone national parks. *Limnology and Oceanography*, **12**, 13–17.

Strijbosch, H. (1979) Habitat selection of amphibians during their aquatic phase. *Oikos*, **33**, 363–72.

Strong, D. R. jr, Simberloff, D., Abele, L. G. and Thistle, A. B. (eds.) (1984) *Ecological Communities: Conceptual Issues and the Evidence.* Princeton University Press, Princeton.

Swales, S. and O'Hara, K. (1980) Instream habitat improvement devices and their use in freshwater fisheries management. *Journal of Environmental Management,* **10**, 167–79.

Swales, S. and O'Hara, K. (1983) A short-term study of the effects of a habitat improvement programme on the distribution and abundance of fish stocks in a small lowland river in Shropshire. *Fisheries Management,* **14**, 135–44.

Talling, J. F. (1951) The element of chance in pond populations. *The Naturalist,* **1951**, 157–70.

Taylor, R. J. (1984) *Predation.* Chapman and Hall, London and New York.

Tervet, D. J. and Harriman, R. (1988) Changes in pH and calcium after selective liming in the catchment of Loch Dee, a sensitive and rapid turnover loch in south-west Scotland. *Aquaculture and Fisheries Management,* **19**, 73–95.

Townsend, C. R., Hildrew, A. G. and Francis, J. E. (1983) Community structure in some southern English streams: the influence of physicochemical factors. *Freshwater Biology,* **13**, 521–44.

Townsend, G. H. (1975) Impact of the Bennett Dam on the Peace-Athabasca Delta. *Journal of the Fisheries Research Board of Canada,* **32**, 171–6.

Van der Velde, G., Meuffels, H. J. G., Heine, M. and Peeters, P. M. (1985) Dolichopodidae (Diptera) of a Nymphaeid dominated system in the Netherlands; species composition, diversity, spatial and temporal distributions. *Aquatic Insects,* **7**, 189–207.

Vangenechten, J. H. D. (1988) Acidification and ecophysiology of freshwater animals. In *Air Pollution and Ecosystems* (ed. P. Mathy). Reidel, Dordrecht.

Vannote, R. L., Minshall, G. W., Cummins, K. W., Sedell, J. R. and Cushing, C. E. (1980) The river continuum concept. *Canadian Journal of Fisheries and Aquatic Sciences,* **37**, 130–7.

Vermeij, G. J. (1982) Unsuccessful predation and evolution. *American Naturalist,* **120**, 701–20.

Vollenweider, R. A. (1968) *Scientific Fundamentals of the Eutrophication of Lakes and Flowing Waters, with Particular Reference to Nitrogen and Phosphorus as Factors in Eutrophication.* Organisation for Economic Co-operation and Development, Paris.

Ward, J. V. (1976) Comparative limnology of differentially regulated sections of a Colorado mountain river. *Archiv für Hydrobiologie,* **78**, 319–42.

Weber, C. A. (1907) *Aufban und vegetation der Moore Norddeutschlands-Beiblatt zu der Botanischen Jahrbuchern,* **90**, 19–34.

Wellburn, A. (1988) *Air Pollution and Acid Rain: the biological impact.* Longman, Singapore.

Westlake, D. F. (1966) The light climate for plants in rivers. In *Light as an Ecological Factor* (eds. R. Bainbridge, G. C. Evans and O. Rackham). British Ecological Society Symposium No. 6. Blackwell, Oxford.

Wetzel, R. G. (1975) *Limnology.* Saunders College Publishing, Philadelphia.

Whitton, B. A. (ed.) (1975) *River Ecology.* Blackwell, Oxford.

Williams, D. D. (1984) The Hyporheic zone as a habitat for aquatic insects and associated arthropods. In *The Ecology of Aquatic Insects* (eds. V. H. Resh and D. M. Rosenberg). Praeger, New York.

Williams, D. D. (1987) *The Ecology of Temporary Water.* Croom Helm, London.

Williams, G. (1980) Swifter flows the river. *Birds,* **8**, 19–22.

Williamson, K. (1971) A bird census of a Dorset dairy farm. *Bird Study,* **18**, 80–96.

Woodiwiss, F. S. (1964) The biological system of stream classification used by the Trent River Board. *Chemistry and Industry,* **11**, 443–7.

Wolf, P. H. (1961) Land drainage and its dangers as experienced in Sweden – II, IV and V. *Salmon and Trout Magazine,* (161), 24–30; (162), 95–100; (163), 145–50.

Wootton, R. J. (1988) The historical ecology of aquatic insects; an overview. *Palaeogeography, Palaeoclimatology, Palaeoecology,* **62**, 477–492.

Wright, J. F., Cameron, A. C., Hiley, P. D. and Berrie, A. D. (1982) Seasonal changes in biomass of macrophytes on shaded and unshaded sections of the River Lambourn, England. *Freshwater Biology,* **12**, 271–83.

Wright, J. F., Moss, D., Armitage, P. D. and Furse, M. T. (1984) A preliminary classification of running water sites in Great Britain based on macro-invertebrate species and the prediction of community type using environmental data. *Freshwater Biology,* **14**, 221–56.

Xuefang, Y. (1983) Possible effects of the proposed eastern transfer route on the fish stock of the principal water bodies along the course, In *Long-Distance Water Transfer: A Chinese Case Study and International Experiences* (eds. A. Biswas, Z. Dakang, J. Nickum and L. Changming). Tycooly International, Dun Laoghaire. pp. 374–88.

Zaret, T. M. (1980) *Predation and Freshwater Communities.* Yale University Press, New Haven.

Zaret, T. M. and Suffern, J. S. (1976) Vertical migration in zooplankton as a predator avoidance mechanism. *Limnology and Oceanography,* **21,** 804–13.

Subject index

Index to Genera and Species

Index of common English names of plants and animals